INTRODUCTION TO
SYSTEMS ENGINEERING FOR THE
INSTRUMENTATION AND CONTROL OF
NUCLEAR FACILITIES

The following States are Members of the International Atomic Energy Agency:

AFGHANISTAN
ALBANIA
ALGERIA
ANGOLA
ANTIGUA AND BARBUDA
ARGENTINA
ARMENIA
AUSTRALIA
AUSTRIA
AZERBAIJAN
BAHAMAS
BAHRAIN
BANGLADESH
BARBADOS
BELARUS
BELGIUM
BELIZE
BENIN
BOLIVIA, PLURINATIONAL
    STATE OF
BOSNIA AND HERZEGOVINA
BOTSWANA
BRAZIL
BRUNEI DARUSSALAM
BULGARIA
BURKINA FASO
BURUNDI
CAMBODIA
CAMEROON
CANADA
CENTRAL AFRICAN
    REPUBLIC
CHAD
CHILE
CHINA
COLOMBIA
COMOROS
CONGO
COSTA RICA
CÔTE D'IVOIRE
CROATIA
CUBA
CYPRUS
CZECH REPUBLIC
DEMOCRATIC REPUBLIC
    OF THE CONGO
DENMARK
DJIBOUTI
DOMINICA
DOMINICAN REPUBLIC
ECUADOR
EGYPT
EL SALVADOR
ERITREA
ESTONIA
ESWATINI
ETHIOPIA
FIJI
FINLAND
FRANCE
GABON
GEORGIA

GERMANY
GHANA
GREECE
GRENADA
GUATEMALA
GUYANA
HAITI
HOLY SEE
HONDURAS
HUNGARY
ICELAND
INDIA
INDONESIA
IRAN, ISLAMIC REPUBLIC OF
IRAQ
IRELAND
ISRAEL
ITALY
JAMAICA
JAPAN
JORDAN
KAZAKHSTAN
KENYA
KOREA, REPUBLIC OF
KUWAIT
KYRGYZSTAN
LAO PEOPLE'S DEMOCRATIC
    REPUBLIC
LATVIA
LEBANON
LESOTHO
LIBERIA
LIBYA
LIECHTENSTEIN
LITHUANIA
LUXEMBOURG
MADAGASCAR
MALAWI
MALAYSIA
MALI
MALTA
MARSHALL ISLANDS
MAURITANIA
MAURITIUS
MEXICO
MONACO
MONGOLIA
MONTENEGRO
MOROCCO
MOZAMBIQUE
MYANMAR
NAMIBIA
NEPAL
NETHERLANDS
NEW ZEALAND
NICARAGUA
NIGER
NIGERIA
NORTH MACEDONIA
NORWAY
OMAN
PAKISTAN

PALAU
PANAMA
PAPUA NEW GUINEA
PARAGUAY
PERU
PHILIPPINES
POLAND
PORTUGAL
QATAR
REPUBLIC OF MOLDOVA
ROMANIA
RUSSIAN FEDERATION
RWANDA
SAINT KITTS AND NEVIS
SAINT LUCIA
SAINT VINCENT AND
    THE GRENADINES
SAMOA
SAN MARINO
SAUDI ARABIA
SENEGAL
SERBIA
SEYCHELLES
SIERRA LEONE
SINGAPORE
SLOVAKIA
SLOVENIA
SOUTH AFRICA
SPAIN
SRI LANKA
SUDAN
SWEDEN
SWITZERLAND
SYRIAN ARAB REPUBLIC
TAJIKISTAN
THAILAND
TOGO
TONGA
TRINIDAD AND TOBAGO
TUNISIA
TÜRKİYE
TURKMENISTAN
UGANDA
UKRAINE
UNITED ARAB EMIRATES
UNITED KINGDOM OF
    GREAT BRITAIN AND
    NORTHERN IRELAND
UNITED REPUBLIC
    OF TANZANIA
UNITED STATES OF AMERICA
URUGUAY
UZBEKISTAN
VANUATU
VENEZUELA, BOLIVARIAN
    REPUBLIC OF
VIET NAM
YEMEN
ZAMBIA
ZIMBABWE

The Agency's Statute was approved on 23 October 1956 by the Conference on the Statute of the IAEA held at United Nations Headquarters, New York; it entered into force on 29 July 1957. The Headquarters of the Agency are situated in Vienna. Its principal objective is "to accelerate and enlarge the contribution of atomic energy to peace, health and prosperity throughout the world".

IAEA NUCLEAR ENERGY SERIES No. NR-T-2.14

INTRODUCTION TO SYSTEMS ENGINEERING FOR THE INSTRUMENTATION AND CONTROL OF NUCLEAR FACILITIES

INTERNATIONAL ATOMIC ENERGY AGENCY
VIENNA, 2022

COPYRIGHT NOTICE

© IAEA, 2022

Printed by the IAEA in Austria
October 2022
STI/PUB/2018

IAEA Library Cataloguing in Publication Data

Names: International Atomic Energy Agency.
Title: Introduction to systems engineering for the instrumentation and control of nuclear facilities / International Atomic Energy Agency.
Description: Vienna : International Atomic Energy Agency, 2022. | Series: IAEA nuclear energy series, ISSN 1995–7807 ; no. NW-T-2.14 | Includes bibliographical references.
Identifiers: IAEAL 22-01521 | ISBN 978–92–0–128522–5 (paperback : alk. paper) | ISBN 978–92–0–128622–2 (pdf) | ISBN 978–92–0–128722–9 (epub)
Subjects: LCSH: Systems engineering. | Systems engineering — Nuclear facilities. | Nuclear facilities. | Nuclear facilities — Engineering — Utilization.
Classification: UDC 62:349.7 | STI/PUB/2018

FOREWORD

The IAEA's statutory role is to "seek to accelerate and enlarge the contribution of atomic energy to peace, health and prosperity throughout the world". Among other functions, the IAEA is authorized to "foster the exchange of scientific and technical information on peaceful uses of atomic energy". One way this is achieved is through a range of technical publications including the IAEA Nuclear Energy Series.

The IAEA Nuclear Energy Series comprises publications designed to further the use of nuclear technologies in support of sustainable development, to advance nuclear science and technology, catalyse innovation and build capacity to support the existing and expanded use of nuclear power and nuclear science applications. The publications include information covering all policy, technological and management aspects of the definition and implementation of activities involving the peaceful use of nuclear technology.

The IAEA safety standards establish fundamental principles, requirements and recommendations to ensure nuclear safety and serve as a global reference for protecting people and the environment from harmful effects of ionizing radiation. When IAEA Nuclear Energy Series publications address safety, it is ensured that the IAEA safety standards are referred to as the current boundary conditions for the application of nuclear technology.

Systems engineering is a holistic, interdisciplinary and cooperative approach to the engineering of large systems such as nuclear power plants, other nuclear facilities and their instrumentation and control (I&C) systems over their life cycles. It is increasingly considered in many industrial sectors to be a necessary means to address the daunting challenges of facing the development and utilization of modern systems caused by ever increasing complexity. The ISO/IEC/IEEE 15288 standard Systems and Software Engineering — System life Cycle Processes was published in 2015 to provide a common process framework.

This publication is an introduction to systems engineering in a nuclear facility and in the context of I&C. Its goal is to assist Member States in understanding the philosophy and methodologies of systems engineering as presented by the ISO/IEC/IEEE 15288 standard and to provide guiding principles for the application of systems engineering to nuclear facilities and their I&C. However, as systems engineering is an extremely broad subject, and as each nuclear facility and organization has specific issues, even in the limited domain of facility I&C, this publication cannot be considered as an implementation guide. Rather, whenever appropriate and possible, it refers to other publications for detailed, practical aspects.

The publication was produced by a committee of international experts and advisors from numerous countries. The IAEA wishes to acknowledge the valuable assistance provided by the contributors and reviewers listed at the end of the publication, especially K. Kolchev (Russian Federation) and T. Nguyen (France), who served as the co-chairs of the authoring group. The IAEA officer responsible for this publication was J. Eiler of the Division of Nuclear Power.

CONTENTS

1. INTRODUCTION

1.1. BACKGROUND

Experience shows that without a rigorous and well organized approach to developing nuclear facilities or nuclear facility systems, including instrumentation and control (I&C), the resulting systems may lack properly defined and traceable requirements or may exhibit unintended and undesirable behaviour which can be potentially unsafe and/or extremely costly. When plant systems become more numerous, more ambitious, often more complex and interdependent, or when innovative features are introduced, adoption of a structured engineering approach becomes even more critical to avoid these situations. Also, nuclear facilities operate in a competitive environment, and meeting tight budgets and schedules is important for them to remain viable. The application of appropriate systems engineering principles can help these facilities increase their viability.

This publication considers the adoption of systems engineering principles in the development of nuclear facilities and their I&C systems that impact on organizational, technical and management processes regardless of their complexity. In this context these principles are applicable to both simple and complex I&C systems.

Every nuclear facility engineering discipline is connected to many others. For example, I&C engineering is linked to disciplines such as fault analysis, heating, ventilation and air conditioning (HVAC) engineering, civil engineering, process engineering, electrical engineering, human factors engineering (HFE), facility layout, etc. Unfortunately, different disciplines often have inadequately connected life cycle processes, different scientific and technical bases, varying perspectives, different methods, diverse constraints, and separate terminologies. Thus, there is often difficulty in understanding what is occurring in another, albeit related, area. A systems engineering approach can help to establish connections between disciplines and determine inputs for, and outputs from, each stage of the system development life cycle, including operation and maintenance (O&M).

Considering these issues, the IAEA Technical Working Group on Nuclear Power Plant Instrumentation and Control (TWG-NPPIC) identified the need for a publication that provides a set of principles for Member States to encourage the adoption and use of systems engineering when developing nuclear facilities and to inform I&C engineers on the need to coordinate with other nuclear facility engineering disciplines. This publication seeks to provide an understanding of systems engineering and how its principles can be applied to a nuclear facility, and to its I&C systems specifically, for the specification of systems requirements, design, implementation and O&M.

1.2. OBJECTIVE

This publication is not intended to be an implementation guide: systems engineering is too broad and projects and organizations are too different, even in the limited area of nuclear facility I&C. Rather, it is an introduction to systems engineering in a nuclear facility and I&C context, taking account of the weaknesses seen in many projects (e.g. insufficient interdisciplinary coordination, lack of rigour in the identification of requirements, and limited use of techniques such as modelling and simulation (M&S)). In particular, it aims at assisting Member States understand the philosophy and methodologies of systems engineering as presented by the ISO/IEC/IEEE 15288 standard, Systems and Software Engineering: System Life Cycle Processes [1], and at providing guiding principles for systems engineering methodologies for nuclear facilities and their I&C system throughout their life cycle. Based on this introduction, more detailed sources can be consulted, such as the Electric Power Research Institute's Digital Engineering Guide: Decision Making Using Systems Engineering [2].

Systems engineering also supports digitization in development and O&M processes. This publication will help Member States transition from 'paper' or 'digital paper' processes to fully digital processes featuring business or engineering process management systems. This improvement may also lead to cost reductions and shorter schedules. Guidance provided here, describing good practices, represents expert opinion but does not constitute recommendations made on the basis of a consensus of Member States.

1.3. SCOPE

Systems engineering principles are not intended to be applied to individual systems in isolation. Indeed, I&C operates within the broader context of the nuclear facility and throughout its operational lifetime. Many disciplines and plant systems apart from I&C have to be considered if systems engineering principles are to be effectively applied. In this way, interactions between I&C systems and non-I&C systems, as well as the overall effects on plant O&M and decommissioning and deconstruction, can be addressed.

This publication provides general definitions and principles for systems engineering which are applicable to the entire nuclear facility. It also provides information on specific aspects of I&C engineering in the framework of nuclear facility engineering, including identification of the interfaces and relevant inputs from and outputs to the environment of the I&C systems being developed.

Also, as I&C includes hardware (sensors, programmable logic controllers, hardwired logic, cabling, supervisory control equipment, mosaic panels, etc.), software (system software, application software, etc.), mathematical models and algorithms, data to support system and subsystem configuration, human–system interfaces (HSIs), etc., it needs to be recognized that I&C engineering is itself composed of more specialized engineering disciplines. Thus, the scope of this publication includes not only the I&C disciplines but also their interactions with other nuclear facility engineering disciplines. Although many of the references provided in this publication focus primarily on nuclear power plants (NPPs), often their principles can also be applied to other nuclear facilities.

This publication is relevant for the following types of projects:

— *New builds.* I&C systems for new nuclear facilities.
— *Modifications.* I&C modifications performed on operating nuclear facilities covering:
  • Large scope modifications affecting multiple I&C systems (and possibly other plant systems) with interconnected functions, plant wide effects and partly common engineering.
  • Small scope modifications affecting only one or very few I&C systems, with isolated functionality and limited effects on the rest of the facility.

The I&C system of a nuclear facility is generally arranged in multiple hierarchical levels: the overall I&C architecture organizes the individual I&C systems of the facility into a structure that meets defence in depth and independence requirements. The individual I&C systems often consist of multiple subsystems (e.g. the redundant divisions of a safety I&C system or the segments of a control system important to production that enable fault tolerance). In the following, the term 'I&C' refers broadly to the complete set of I&C systems of the facility, whereas the term 'I&C system' refers to a specific I&C system.

1.4. STRUCTURE

This publication is organized into six sections, including Section 1, and an Appendix. Section 2 defines systems engineering, building mostly on the ISO/IEC/IEEE 15288 [1] standard, and explains why it is important for nuclear facilities and their I&C systems. Section 3 introduces the major processes, including those applicable to organizational, technical, management and regulatory activities, used in systems engineering and refers to various guidance documents that can be used to implement them. Section 4 introduces methodologies such as modelling, justification framework and knowledge management that support the systems engineering processes. Section 5 covers tools that could be used for systems engineering. Section 6 contains a summary. The appendix provides examples of selected processes for the development of I&C systems in nuclear facilities.

2. OVERVIEW OF SYSTEMS ENGINEERING

2.1. INTRODUCTION

Systems engineering is a broad concept that has evolved into both an engineering approach and an engineering discipline in itself. It is founded on the realization that systems are composed of increasing numbers of interdependent elements, that their interactions become ever more numerous, and that errors could have critical or even unacceptable consequences. Thus, the combinations and conditions that need to be considered can increase such that large systems become extremely difficult to comprehend and require the contribution of a wide range of stakeholders and engineering disciplines and a large number of persons and teams.

Systems engineering is intended to bring comprehension to large and complex systems, and to organize the cooperation of all those involved. Indeed, although the wordings of the ISO/IEC/IEEE 15288 [1] standard and of the International Council on Systems Engineering (INCOSE) definitions differ, they both view systems engineering as a holistic, interdisciplinary and cooperative approach to the engineering of systems over their entire life cycles. As they promote 'systems thinking', i.e. the idea that everything is a system, they also apply it to the management of that engineering.

The core of systems engineering is based on a set of well focused and interrelated processes, each using inputs from the others and in turn providing them with feedback in an iterative manner. Thus, the full engineering process integrates more refined information and decisions in each iteration until a solution emerges as a synthesis of the full body of requirements. This can help track the necessary trade-offs between conflicting objectives and constraints, and the rationales for decisions and changes made at any point in the life cycle.

Indeed, support for the full life cycle of a system, from inception through disposal, is a key aspect of systems engineering. There are numerous life cycle models which will be discussed elsewhere in this publication. Such models now tend to be comprehensive and take account not only of systems themselves, but also of corporate organizations, societal goals and other socioeconomic factors. However, some are simplified or truncated to focus on specific areas of need or on limited parts of the life cycle (e.g. projects) while still using the core systems engineering model.

With the progressive expansion of stakeholder needs, first into system requirements and then into increasingly detailed solutions, and with the integration of pre-existing solutions, elements and products, systems engineering allows the engineering to be organized into a continuous integration process that can be diagnosed for flaws and completed with confidence. In particular, it helps resolve design conflicts and balance stakeholder needs with acceptable risk and low life cycle cost.

While systems engineering can be used for any or all systems in a nuclear facility, the emphasis in this publication is on I&C, including the new digital and software based technologies that are prevalent today. To be clear, I&C can be an embedded part of a single fluid/mechanical/electrical plant system, or I&C functions can be gathered into a system that monitors and controls multiple plant systems. In all cases, I&C systems can be visualized as the 'controlling system' and the fluid, mechanical or electrical systems or components as the 'controlled system'. This will assist in compartmentalizing the use of systems engineering to I&C within nuclear facilities.

2.2. WHY SYSTEMS ENGINEERING IS IMPORTANT FOR NUCLEAR FACILITIES

Many factors contribute to make systems engineering important for nuclear facilities:

— Numerous stakeholders, disciplines and teams;
— Need to be competitive and to innovate;
— Need to justify safety and security;
— Extremely long life cycles.

In the engineering of a nuclear facility and its I&C systems, it is necessary to coordinate numerous stakeholders, including owners, designers, suppliers, builders, operators, regulators, societal representatives, grid managers and unions, as well as numerous engineering disciplines such as I&C, safety, security, probabilistic analysis, process design, electrical design, operation, maintenance, construction and hazard analysis. Each tends to have its own place in the facility and its own engineering culture, methods and tools. Thus, communication and coordination are not easy tasks, and experience in all industrial sectors shows that rigorous and effective approaches are needed. Systems engineering provides a framework to bring together the various stakeholders, design, maintenance and operational resources into a team based approach to system implementation. This can result in faster, safer and more efficient decision making and can avoid late design changes, which translates to shorter and more successful projects and more efficient operation, as has been demonstrated in the transportation and process industries.

One of the main challenges nuclear facilities are facing today is economic: they need to be competitive with respect to other sources of energy which require facilities that are increasingly faster, cheaper and easier to construct and operate. Traditional evolutionary engineering approaches, where new facilities are based on proven solutions with only limited changes, do not always provide a satisfactory answer. With a better understanding of systems as a whole and of the possible (otherwise unforeseen) interactions between components and between disciplines, systems engineering can enable efficient, safe and secure implementation of innovations in design and operation.

Licensing is a significant part of nuclear facilities and of the related I&C engineering. Well applied and rigorous systems engineering approaches can bring clarity and completeness, and thus facilitate understanding between licensors and licensees.

The life cycle of nuclear facilities is such that engineers who initiated such a construction project will not necessarily complete its design and construction and support its operation, its upgrades, and ultimately, its decommissioning and deconstruction. Systems engineering addresses this problem through organized and systematic information and knowledge management processes. This allows safety, security and engineering knowledge on the facility and its systems to be conveyed to the future workforce in an effective manner.

In and of itself, systems engineering is not a 'magic bullet' but, when properly tailored to the specifics of a system of interest and its project(s), it provides a compelling framework for the improvement of current engineering methods.

2.3. SYSTEMS ENGINEERING FOR NUCLEAR INSTRUMENTATION AND CONTROL

I&C, which is often viewed as a facility's central neural system, is among the plant systems most affected by greater complexity. Many I&C systems in new or upgraded plants are digital, and their size and complexity raise issues that analogue systems do not face in terms of safety, security, human factors, HVAC, power supply, equipment qualification, rapid obsolescence and knowledge management, to name just a few. Some of these issues even place constraints on systems.

In addition to technological complexity, functional complexity is also increasing. Indeed, the flexibility allowed by digital technologies has led to ever higher functional requirements for I&C. As a consequence, the associated risk of requirement specification errors is an increasing concern, in particular due to the fact that requirement specification is at the beginning of the I&C life cycle, and that any error at that stage could be revealed only late in the I&C development process, or worse, during operation (with possible significant consequences on schedule and cost). To prevent such errors, effective coordination is needed with process engineering, HFE, hazard and risk analysis, safety engineering, security engineering and operations planners.

Lastly, recent experience shows that I&C now represents a significant part of the cost and engineering of a nuclear facility and, in some Member States, of the licensing difficulties and uncertainties. Thus, its cost effectiveness, its on-schedule implementation and its conformance with respect to regulatory requirements are vital to the success of a nuclear facility.

Systems engineering has been used to address similar challenges in a variety of other industries. It enables a holistic and integrated approach for the I&C systems of a nuclear facility, and its principles can be applied to integrate, evaluate and balance the constraints and contributions from all other engineering disciplines concerned, so as to produce a coherent whole that is not dominated by any single discipline.

An example highlighting the importance of systems engineering for I&C is presented here. It is based on a civil aviation event: the crash of Lion Air B737 MAX. The summary provided here is taken from the official final report of the Indonesian Transportation Safety Board [3]. Civil aviation is also a high safety industrial sector, with extensive experience in operation (on average, a take-off occurs every second all year round) and in-depth investigation and public reporting in the event of an accident.

The blame for this accident has often been placed on an I&C system, the manoeuvring characteristics augmentation system (MCAS). However, the official investigation report offers a nuanced conclusion:

— Early functional hazards analysis considered two MCAS malfunctions:
  • Spurious MCAS operation up to its maximum authority (0.6 degrees);
  • Spurious MCAS operation equivalent to a 3 s stabilizer trim runaway.
— The report classified the consequences as 'major' (a relatively low safety importance, comparable to category C and class 3 of IEC 61226 [4]) on the assumption that pilots could reliably correct MCAS spurious actions within a delay of 3 s. This meant that extremely rigorous verification and validation (V&V), fault tolerance and extensive failure modes and effects analysis were not required.
— Engineering decisions made by other teams and disciplines later in the project disproved the assumption and will ideally have led to a reclassification of 'hazardous' (comparable to category B and class 2 of IEC 61226 [4]):
  • Pilots were not informed of the existence of the MCAS. The Lion Air crew did not react to MCAS actions but to the increasing force on the control column;
  • MCAS authority was raised to 2.5 degrees to address particular flight conditions;
  • MCAS performed multiple actuations in quick sequence for a single event.
— Thus, a significant part of the blame needs to be placed on insufficiently rigorous systems engineering and inadequate coordination between teams and disciplines along the project.

2.4. THE ISO/IEC/IEEE 15288 STANDARD

ISO/IEC/IEEE 15288 [1] is an international standard that establishes a common framework for describing, understanding and applying systems engineering principles to the life cycle of systems. Two major concepts are vital to understand the standard: process and the life cycle model. A process is a set of interrelated or interacting activities, while a life cycle model organizes the processes and activities concerned with the life cycle into stages (see Glossary). Each process can be used whenever needed, as specified by the life cycle model. It is worthwhile noting that, though 'system life cycle' is emphasized in its title, ISO/IEC/IEEE 15288 [1] does not prescribe a specific life cycle model. Instead, it has a specific process for defining, approving and managing a life cycle model or models. For nuclear facility I&C

systems, the life cycle model presented in Ref. [5] can be used. The processes defined in the ISO/IEC/IEEE 15288 standard [1] can then be tailored to match the specific needs of the system under consideration.

The standard identifies four process groups, as summarized in Fig. 1:

(1) Technical processes are focused on the system of interest (e.g. an I&C system, the complete I&C of a nuclear facility, or the facility itself) and enable coordination between all concerned engineers, engineering disciplines and system stakeholders.

(2) Technical management processes are focused on the management of the resources and assets necessary for projects and activities, and apply throughout an organization.

(3) Agreement processes are focused on relations between organizations, e.g. between purchasers and vendors.

(4) Organizational project enabling processes are focused on the organization's capability to carry out projects.

Figure 2 shows a simplified but typical arrangement of these processes for an individual nuclear facility I&C system with the addition of a regulatory/licensing process, which is of particular importance to systems important to safety.

In Section 3, processes of importance to nuclear I&C applications are described in detail. The following processes are not discussed further in this publication, since no additional information beyond what is provided in ISO/IEC/IEEE 15288 [1] is needed:

— Organizational project enabling processes not elaborated:
  • Portfolio management;
  • Human resource management;
  • Quality management.
— Technical management processes not elaborated:
  • Project assessment and control;
  • Decision management;
  • Risk management;
  • Measurement;
  • Quality assurance.
— Agreement processes not elaborated:
  • Acquisition;
  • Supply.

As the standard is a general top level document, it does not address particular industrial sectors, types of systems or types of projects, and its descriptions, requirements and recommendations are generic. While it is a good basis for the application of systems engineering to nuclear facilities and their I&C systems, it needs to be tailored to the specific needs of each system or project. In particular, some proposed processes might not be relevant to the system or project, whereas processes important for that system or project might not be mentioned by the standard. Also, certain processes may come under different designations.

More guidance on the application of this standard can be found in the INCOSE Systems Engineering Handbook [6]. Section 3 of this publication provides specific guidance on the systems engineering processes and activities for nuclear facility I&C systems.

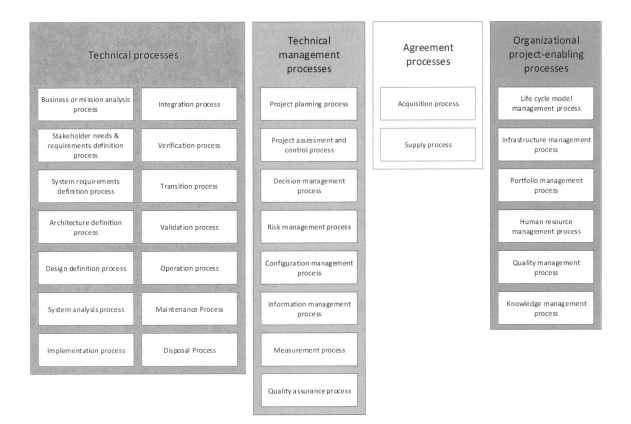

FIG. 1. Grouping of ISO/IEC/IEEE 15288 systems engineering processes.

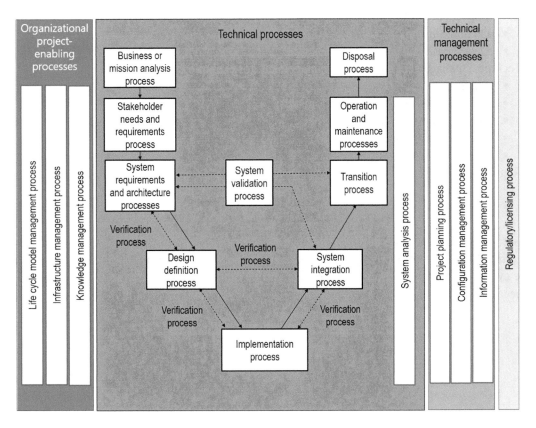

FIG. 2. A systems engineering process model for an individual nuclear facility I&C system.

2.5. INTERLINKAGE OF INSTRUMENTATION AND CONTROL SYSTEMS WITH OTHER DISCIPLINES

The complexity of nuclear facility I&C systems has been outlined in Section 2.3; many disciplines will be involved in I&C engineering projects. One key objective of systems engineering is to ensure there is appropriate communication and coordination, in particular so that I&C engineering personnel have the information needed from the other engineering disciplines and experts in the other disciplines are informed of the needs I&C engineering places on them. Under the umbrella of systems engineering methodologies, it is a good practice to identify these other disciplines explicitly and systematically since organizations in these areas may be in charge of, or associated with, the entities responsible for I&C system. They may also be in charge of particular stages of the plant or the I&C life cycle.

Examples of the potential interfacing disciplines include:

— Plant and plant systems design, for example:
  - Plant process engineering;
  - Plant layout;
  - HFE;
  - Electrical power supply and electromagnetic compatibility;
  - HVAC.
— Safety analysis and performance evaluation:
  - Hazards and risks analysis;
  - Deterministic safety analysis;
  - Probabilistic safety assessment, including human reliability analysis;
  - Equipment qualification.
— Plant O&M:
  - Concept of O&M;
  - Commissioning;
  - Decommissioning and deconstruction.
— Technical management and project management, for example:
  - Costing;
  - Construction and construction logistics;
  - Project management and planning;
  - Configuration management (CM);
  - Documentation management.
— Regulatory area/licensing:
  - Licensing;
  - Life extension application.

Section 3 provides more detailed guidance on this subject, with suggestions on which bodies in engineering disciplines are likely to need to interact with I&C engineering entities at a given stage of the life cycle, what types of information are likely to be required and what types of decisions are likely to be necessary.

2.6. CONCURRENCY, ITERATION AND RECURSION

There will not be a sequential single pass through the processes shown in Fig. 2 and Section 3. Concurrency, iteration and recursion are three major ways of integrating different life cycle processes:

— With concurrency, some processes are performed in parallel, to reduce the time needed but also to facilitate the necessary cross-cutting interactions between processes. This is sometimes called concurrent engineering, or integrated product development.
— With iteration, the same process or set of processes are repeated at the same level of system hierarchy to arrive at an effective solution.
— With recursion, the same process, or set of processes, are applied at embedded levels of the system hierarchy on parts that are also considered systems. The outputs from one level become inputs to the next embedded level. For example, in the case of I&C, outputs from the architecture definition process for the overall I&C (e.g. defence in depth concepts, diversity and redundancy) become inputs to the system requirements definition processes for each I&C system.

2.7. CHALLENGES IN THE APPLICATION OF SYSTEMS ENGINEERING

Although systems engineering is essential for the success of nuclear facilities and nuclear facility I&C projects, care is needed when introducing it in an organization or in an ongoing project; several issues need to be considered.

First, systems engineering is a very general approach with a very large range of systems and projects for different industrial sectors and applications, all of which have different constraints and practices. Systems engineering guidance, training, training material and support services may be very generic and not tailored to the needs and context of a particular application, most particularly of a nuclear facility. They are sometimes abstract and not directly related to systems engineering's proclaimed goals of rigour and effectiveness for answering stakeholder needs. An inappropriate application of systems engineering guidance may lead to misunderstanding, unnecessary activities, costs and delays, inadequate outcomes and, in the end, disillusionment. It is necessary to determine, for the system and the organizations concerned, how the guidance needs to be translated into specific, efficient and practical processes recognizing organizational and cultural specificities.

Second, since introducing systems engineering in an organization takes time (for training, for introducing new processes or modifying existing ones, to create the necessary engineering artefacts, etc.), it is necessary to ensure that ongoing processes are not disrupted to the point of causing unacceptable delays and costs. It is also necessary to consider that multiple organizations are concerned (e.g. suppliers and service providers) with different engineering processes and levels of mastery of systems engineering. Introducing full-fledged systems engineering to a pre-existing system or to an ongoing project is not always the most optimal course of action, and sometimes more graded approaches involving only selected subsystems, teams and life cycle activities might be preferable and provide useful lessons before widespread application.

Third, introducing systems engineering is a challenging endeavour requiring organization-wide decisions, investments and infrastructure, and active, continuous, but also enlightened, support from top level management. Workforce development and adherence are also critical factors without which there can be no real success.

2.8. GRADED APPLICATION OF SYSTEMS ENGINEERING PRINCIPLES

The large number of processes proposed by ISO/IEC/IEEE 15288 and their extensive discussion in this publication may seem daunting and, considering the challenges, it might be considered that for small, limited projects the effort is not worth undertaking, or that for large, extensive projects the learning curve is too steep. This is not desirable.

The processes of ISO/IEC/IEEE 15288 are classical: many technical and technical management processes are extensively addressed in existing standards, regulations and guidelines and most, if not all, are actually implemented in nuclear facilities and facility I&C projects. Systems engineering is, at its core, a state of mind and a culture that aims at enhancing the effectiveness of these processes with:

— A clear understanding of the key objectives of the system or project (which is the purpose of the business or mission analysis process (see Section 3.2.1));
— Deliberate, active, improved coordination and cooperation of concerned stakeholders, engineering disciplines and teams, as discussed in Sections 2.5 and 3.2.2;
— Explicit consideration of the needs and constraints of processes and activities that will be implemented later in the system life cycle;
— Systematic characterization and examination of the environment, and situations the system or its constituents will face during their life cycle.

In this context, enhanced effectiveness means:

— Focus on the 'reason of being' of the system or project, making sure that its key objectives are satisfied and that it is not jeopardized by features that are not necessary.
— Fewer errors revealed late in the system life cycle, as such errors tend to require extensive, costly and time consuming corrections.
— Better solutions, as optimizations made separately by each team and discipline do not result in a global optimum. Also, last minute changes to correct errors revealed late in the life cycle are often less than optimal.

With this in mind, each project may decide, in the framework of the project planning process (see Section 3.3.1), which parts of the system, which processes of the project and to what extent systems engineering principles are to be applied. Organizations not familiar with them may decide to apply them first on some limited parts of the system, or on small scale, limited projects to gain experience and train personnel before more widespread application. In the case of a small scale project, one may decide to focus effort on processes critical for the project. However, it is preferable not to dismiss any process a priori, and instead carefully consider each one. Indeed, one needs to keep in mind that even seemingly limited upgrade or modification projects may have large, unacceptable consequences.

The Cranbrook manoeuvre case presented in Section 4.4.4 is a typical example. Ensuring that thrust reversers are deenergized when the aircraft is airborne was technically a relatively limited modification, but insufficient coordination with other engineering disciplines led to an incomplete examination of the situations the aircraft may face, which significantly contributed to the catastrophic accident.

3. SYSTEMS ENGINEERING PROCESSES

This section explains how systems engineering processes can be applied to a nuclear facility in general and to facility I&C systems specifically.

3.1. ORGANIZATIONAL PROJECT-ENABLING PROCESSES

The organizational project-enabling processes establish the environment in which projects are conducted. Being at a strategic level of the organization's management, they have a key role for the successful realization of projects. In these processes, the organization:

— Initiates, modifies and terminates projects;
— Chooses, modifies and applies the life cycle models and corresponding processes;
— Provides the required material, informational, human and financial resources;
— Sets and monitors the quality management measures for enabling projects to meet the needs and expectations of the interested parties.

For the purpose of this publication, the following organizational project-enabling processes from Fig. 1 are of the most interest:

— Life cycle model management process;
— Infrastructure management process;
— Knowledge management process.

3.1.1. Life cycle model management process

3.1.1.1. Definition of, and general information on, the life cycle model management process

The purpose of this process is to define and maintain the life cycle model applied to a system. Such a model is expressed in terms of stages, milestones, processes and procedures, which it organizes into an integrated whole. An organization may develop a base life cycle framework consistent with its policies, objectives and resources, and then refine, adapt and improve it for individual systems.

The succession of stages and the milestones of a life cycle model describe progress in the engineering and life of the system. The model identifies the inputs and outputs of each stage and the conditions for moving from one stage to the next, or for achieving a milestone covering the data exchange between internal and external stakeholders.

The model selects the engineering processes necessary to the system with respect to stages and milestones. They include all or a selection of ISO/IEC/IEEE 15288 [1] processes, but may also include specific processes. Technical processes as defined by ISO/IEC/IEEE 15288 [1] have a key role, while the other processes are present at all stages (see Fig. 2).

Each process is performed by the application of a number of well defined, targeted procedures. Figure 3 represents an example of a five stage system development process model. It integrates multiple design aspects of a system's development, which leads to the development of an integrated system that is tested, verified and ready for installation and operation at the facility. This model also applies a set of common support processes to each of the individual development subprocesses. Other models may include additional life cycle phases, such as O&M. By organizing the different aspects of development in this manner, the model provides direct integration paths between development subprocesses, which facilitates the incorporation of requirements in a holistic manner. Such an approach supports early identification and resolution of conflicting or competing requirements. As an example, a security related requirement that restricts data flow between components of a system may conflict with an I&C related requirement that calls for open transfer of information between these components. By identifying these requirements in parallel and prior to performing system integration activities, the developer can build a communication scheme that can satisfy both requirements.

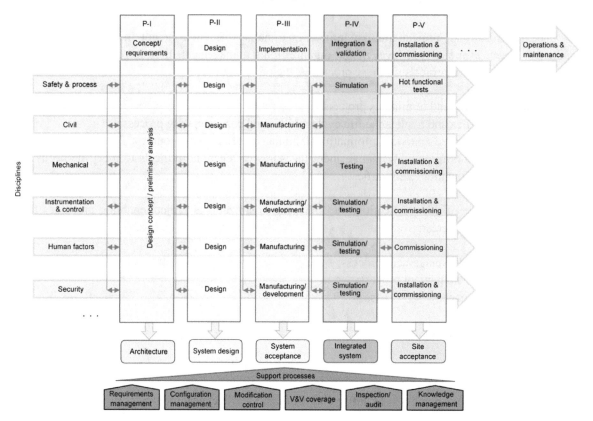

FIG. 3. Example of a process life cycle model with interaction between disciplines.

3.1.1.2. Nuclear facility specific information

For systems as highly regulated as nuclear facilities, it is worthwhile to include a regulatory and licensing process in their life cycle model. The model also needs to be in line with the applicable national regulations and international standards.

A nuclear facility is a very large and complex system: as design progresses, it needs to be divided into subsystems, which are themselves large and complex systems that need to be divided into smaller subsystems. This subdivision may be repeated iteratively and recursively as far as necessary, resulting in what is often called a system breakdown structure. A breakdown structure element may be a product, data, service or any combination thereof. A breakdown structure can also provide a framework for detailed cost estimation and control along with support for schedule development and control. As each element in the breakdown structure may have its own life cycle model, the life cycle model of a composite element integrates and coordinates the life cycle models of its constituents.

Usually, multiple organizations contribute to the engineering of a nuclear facility. When a subsystem is subcontracted or purchased as an off the shelf product, the concerned organizations need to agree on what parts of the subsystem life cycle model need to be integrated and coordinated with those of the facility.

As a nuclear facility breakdown structure includes a very large number of elements, many of which are subcontracted or purchased, there is a strong need for rigorous tools to maintain system information and data exchange between participants. The use of architectural and product breakdown models is also discussed in Section 4.4.5.

Because it is a key element in the engineering of a hugely expensive and very strategic system, the life cycle model management process needs strong support and control from management.

3.1.1.3. Information specific to nuclear facility instrumentation and control systems

Given the scale and importance of a nuclear facility I&C system, the following considerations apply for effective life cycle models:

— The I&C life cycle model integrates and coordinates the life cycles of the overall I&C architecture and of each of the individual I&C systems, paying particular attention to safety aspects (referred to as safety life cycles). Similarly, the life cycle model of an individual I&C system integrates and coordinates the life cycles of its subsystems, hardware components and software components.
— These life cycles need to be aligned with the life cycle of the entire nuclear facility. It is necessary to identify interfaces with other disciplines. Interaction with them and timely information exchange are defined in the I&C life cycle and life cycles of other plant systems.
— Figure 4 shows a typical safety life cycle for an NPP's I&C architecture and its individual I&C systems, together with associated activities and interfaces with HFEs and computer security programmes [5].
— Important technical processes aspects can be found in IAEA Safety Standards Series No. SSG-39 [5], and an I&C top–down design and development approach is described in IEC 61513 [7].
— Within the framework of an I&C project, one of the most important tasks is to identify the scope and the key requirements of the project.

3.1.2. Infrastructure management process

3.1.2.1. Definition and general information

This process specifies, provides and maintains the facilities, tools, technical infrastructure and services that support the other engineering processes during the life cycle, in particular the technical processes. IAEA Safety Standards Series No. SSG-39 [5] provides guidance for the management of system activities, including managing the infrastructure.

3.1.2.2. Information specific to nuclear facilities

It is a good practice to support the various technical processes for nuclear facility engineering with adequate facilities, tools and services (infrastructure). Section 4 provides information on supporting methodologies that can be used to support the infrastructure management process. It is also a good practice to identify the infrastructure needs early in the life cycle so that the necessary facilities, tools and trained personnel are available when needed. Particular attention needs to be given to the long term availability of the infrastructure and its components, since they may be needed for the lifetime of the facility.

The infrastructure management process may need to interface with:

— The configuration management process, since a significant part of what is produced or used by the infrastructure may also need to be managed as configuration items.
— The knowledge management process, to ensure adequate training and maintenance of competences related to the constituents of the infrastructure and the associated methods. Also, to ensure that adequate knowledge regarding the system is maintained during its life cycle.
— The agreement processes, when engineering is distributed among several organizations (typically between the main engineering organization, contractors, equipment vendors and assessors). In some cases, an organization could provide parts of the project infrastructure. In others, the infrastructures of organizations might need to interact with one another.

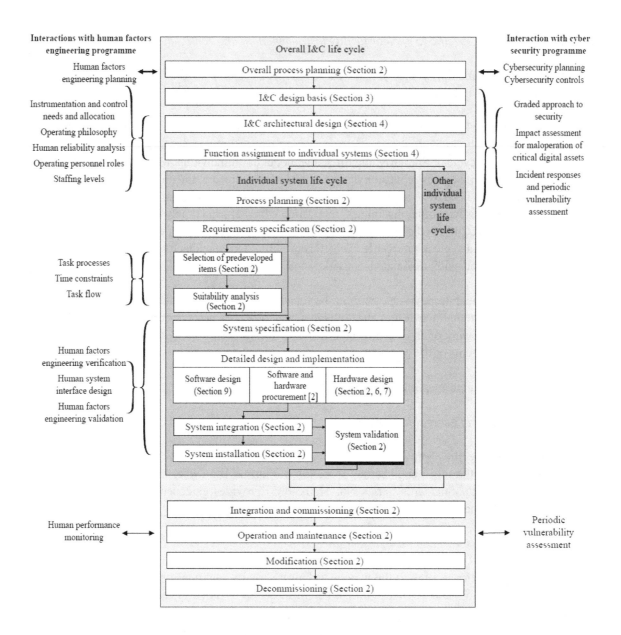

FIG. 4. I&C development life cycle activities and interfaces [5].

3.1.2.3. Information specific to nuclear facility instrumentation and control systems

For I&C systems important to safety, the requirements and recommendations on infrastructure given in the following documents would apply:

— IAEA Safety Standards Series No. SSG-39 [5] (I&C design safety);
— IAEA Safety Standards Series No. SSR-2/1 (Rev. 1) [8] (proven engineering practices and design rules);
— IAEA Nuclear Security Series No. 17-T (Rev. 1) [9] (computer security);
— IAEA Nuclear Security Series No. 33-T [10] (computer security);
— IEC 60880 [11] (software aspects) or IEEE 7-4.3.2 [12] (digital devices in safety systems);
— IEC 60987 [13] (hardware aspects);
— IEC 61508 [14] (functional safety);
— IEC 61513 [7] (safety) or IEEE 603 [15] (safety systems);

— IEC 62138 [16] (software aspects);
— IEC 62566 [17] (hardware description language (HDL) aspects);
— IEC 62645 [18] (computer security).

3.1.3. Knowledge management process

3.1.3.1. Definition and general information

This process ensures that during the life cycle appropriate knowledge is available when needed by the different engineering processes. Knowledge management activities provide a means of developing and acquiring knowledge assets needed for successful system development, implementation and operation.

3.1.3.2. Information specific to nuclear facilities

Different kinds of knowledge may need different approaches:

— Specific and technological knowledge (i.e. knowledge regarding available technologies and products) tend to be generic and not facility specific.
— Operational knowledge (how to do, how to recognize, how to diagnose, etc.) is facility specific. It includes knowledge of the design, construction and operation of the nuclear facility and its systems.
— The knowledge of the design and operation of the facility and its systems, the safety functions and security requirements and the rationales behind the solutions chosen, is facility specific. Documentation provides information, but that is not sufficient: effective knowledge is based on the understanding and experience of that information. Section 4.7 provides more information on this topic and Section 4.5 provides information on the establishment of a justification framework.

Education and training (at the facility itself or with simulators) can provide basic disciplinary, technological and operational knowledge, but expertise is generally obtained only through practice and experience. The knowledge management process serves most of the other processes, but also needs appropriate infrastructure support (e.g. training material, simulators or knowledge repositories).

3.1.3.3. Information specific to nuclear facility I&C systems

Knowledge regarding facility I&C systems may be divided into two main parts: knowledge specific to I&C, and knowledge of the I&C within the framework of the facility. Note that the I&C system of a nuclear facility has knowledge of its own:

— Disciplinary and technological knowledge for I&C is dependent on the technologies chosen. For example, disciplinary knowledge of analogue I&C systems is not the same as for software based I&C systems, which in turn is not the knowledge for field programmable gate array based I&C systems. As technological and regulatory changes occur relatively quickly compared with the lifetime of an I&C system and of the plant, specific technical and agreement measures need to be taken to ensure that even after several decades, the associated parts of the infrastructure (methods, tools, languages, libraries, etc.) can still be easily used maintenance or modification.
— Operational knowledge, knowing how the I&C system interfaces with the nuclear facility and its functional contribution to the operation of the facility.
— Engineering, safety and security knowledge for digital I&C systems needs particular effort: documenting requirements and solutions is often not sufficient. It is worthwhile to document rationales and assumptions and to organize the vast amount of individual knowledge elements into a structured and logical whole.

On the other hand, I&C plays an important role in many aspects of the facility, to the point that it has sometimes been likened to a 'central nervous system' (see Ref. [19]). It is therefore essential that the knowledge management process explicitly addresses the role and effects of the I&C system in the full operational context of the facility: with human operators and other plant systems; with support systems; and with respect to the different plant and plant system states.

3.2. TECHNICAL PROCESSES

3.2.1. Business or mission analysis process

3.2.1.1. Definition and general information

This process defines the problems to be addressed and/or the opportunities to be realized, specifies the main objectives of the system or project, characterizes the solution space and identifies classes of potential solutions. All of these constitute the rationale of the system or project. Plant and I&C engineers need to keep the rationale in mind when looking for, or choosing between, technical solutions for each of the other technical processes.

3.2.1.2. Information specific to nuclear facilities

The business or mission analysis process for nuclear facility or a series of such facilities would typically identify market, economic, societal and technological needs and opportunities, and determine top level objectives such as:

— Leveraged cost of electricity targets.
— Availability targets.
— Safety targets.
— Security targets.
— Manoeuvrability targets (i.e. the ability to adapt plant production to calls for power).
— Siting(s), which may determine:
  • Which safety and security regulations and standards need to be applied;
  • Environmental conditions and issues;
  • Logistics issues for construction and operation;
  • Power grid insertion constraints.
— Services to be provided, for training, operation, maintenance, outages, renovation.

The results of this process may also include a number of top level strategic concepts defining the main characteristics of the nuclear facility or of the series as a whole. One such is a passivity concept (i.e. use of safety systems that need little or no external power and control) and another is a reference design (a nuclear facility is usually not designed completely from scratch).

3.2.1.3. Information specific to nuclear facility instrumentation and control

As this process is often strategic, its outputs are typically provided to the I&C engineers. However, if they do not directly participate in the process, it is essential that they inform those in charge of the opportunities offered by modern I&C technologies, but also of the challenges, needs, limitations and constraints (considering in particular the evolution of safety and computer security regulatory and standard requirements).

3.2.2. Process for defining stakeholder needs and requirements

3.2.2.1. Definition and general information

A stakeholder is an individual or organization with a legitimate interest in the system. Success depends on the system meeting the needs of its stakeholders throughout its life cycle. The first objective of this process is to identify these stakeholders and determine their needs, grading them along an urgency scale, from 'need to have' demands to 'nice to have' desires. A second objective is, given the outcomes of the business or mission analysis process, to translate these needs into stakeholder requirements.

A clear distinction needs to be made between stakeholder needs and stakeholder requirements. Stakeholder needs are often stated in imprecise and ambiguous forms, and sometimes in terms of solutions rather than actual needs. Some may be unachievable by any realistic system, and different stakeholders may view the system from very different standpoints and express antagonistic or even contradictory needs. Stakeholder requirements are precise statements that specify the actual targets of the system and its engineering: they resolve ambiguities and make necessary trade-offs to resolve feasibility issues and contradictions, taking account of the fact that different needs may apply at different life cycle stages.

Since stakeholder requirements are the primary basis for the system requirements definition process and the validation process, it is important that stakeholders confirm that their demands are correctly captured and their desires are addressed to an acceptable degree, even though it is generally very difficult or even impossible to completely satisfy all stakeholders.

3.2.2.2. Information specific to nuclear facilities

Nuclear facility stakeholders include the facility owner(s), operators, managers of the electrical grid, regulators, local authorities, vendors, facility personnel and society at large, to name just a few. A nuclear facility is a complex system with a long lifetime; thus, a wide variety of stakeholders may be involved. During the facility's life cycle, different stakeholders will assume varying importance. When a nuclear programme is under discussion, national stakeholders tend to be more important, whereas once sites have been identified, local stakeholders become a primary focus.

Stakeholders vary from country to country, and the levels and types of engagement can differ. Each stakeholder group has specific information needs and expectations, which may be addressed in different ways depending on the stakeholder profile and the issue under consideration. It is crucial to fully understand stakeholders in terms of their self-stated (or underlying) purpose and their interest or concerns (not always explicitly expressed) [20].

3.2.2.3. Information specific to nuclear facility instrumentation and control systems

For I&C systems, the process for defining stakeholder needs and requirements has two sides:

(1) The first identifies the stakeholders expressing needs and constraints to be addressed by I&C systems, such as plant process engineers, plant systems engineers, operators, maintenance staff, regulatory bodies, and disciplines such as safety, computer security, HFE, equipment qualification and licensing.
(2) The second identifies the stakeholders who are to address the needs and constraints of I&C, such as plant architects, plant layout architects, power supply engineers, HVAC engineers, I&C suppliers, system integrators, operators and maintenance staff.

There may be stakeholders on both sides. Proactive approaches are generally needed to determine needs and derive the requirements acceptable to the parties concerned. Such discussions take place during systems meetings. They can be arranged according to several levels of design. For example:

— Between nuclear facility design and overall I&C design. Different disciplines provide requirements to I&C, but sometimes in the form of solutions (e.g. assuming a specific technology). Systems meetings may help make sure that the real needs of the process are identified.
— Between overall I&C and individual I&C systems design. Designers of the overall I&C system may specify requirements for individual I&C systems that cannot be fulfilled with the technology available. Systems meetings may help all parties come to an agreement on achievable requirements.

3.2.3. Process for defining system requirements

3.2.3.1. Definition and general information

The purpose of this process is to transform the stakeholder requirements, which are often expressed in terms of overall goals, into an organized set of practical and verifiable system requirements providing a technical view of an operable solution. These requirements may also identify and address design, implementation and operational constraints not necessarily mentioned by stakeholders.

It is a good practice to ensure traceability between stakeholder needs and requirements.

3.2.3.2. Information specific to nuclear facilities

As nuclear facilities are large and complex systems, their subsystems (and the subsystems of their subsystems) may themselves be full-fledged systems. This facility specific information also applies in large part to subsystems, including the overall I&C and the individual I&C systems.

As system requirements are the foundation for many technical processes, it is essential that they are of the highest quality. They need to be:

— Complete. Activities related to requirements are often performed throughout a system life cycle to determine the degree of completeness that is achieved. Even though they have to remain at a high level, they not only need to address the full set of specified stakeholder requirements, they also need to take account of the many different situations the facility will face along its lifetime. Some situations are determined by the life cycle model, such as construction on-site, commissioning, active operation, outage, renovation or decommissioning and deconstruction. Others are determined by normal operational states (from shutdown without fuel to full power), by abnormal states (failure or accident conditions), by normal and abnormal external events and conditions and by the operational goals set by operators at any given instant (in the example in Section 4.4.4, the pilots switched from goal 'landing' to goal 'take-off'). Lastly, some engineering disciplines not strategic enough to be considered as stakeholders may need to be involved and may have their say in the process.
— Adequate for all the situations for which they are specified. In the example in Section 4.4.4, the requirement to disengage thrust reversers while airborne was adequate in most situations, but not when pilots abruptly switch from the goal of 'landing' (where the aircraft is configured for landing and reversers are deployed) to 'take-off' (where it was implicitly and incorrectly assumed that the aircraft would be configured for take-off and reversers were fully stowed).
— Unambiguous (i.e. they cannot be interpreted differently by concerned engineers). They will ideally also be verifiable (i.e. they must have clear satisfaction criteria), feasible and consistent (i.e. free of contradictions).

Thus, it is preferable to:

— Have a comprehensive list of topics (e.g. power production, availability, safety, security, environment, finance, economy) and aspects (e.g. functional, performance, process, non-functional and interface aspects) to be addressed.
— Avoid making or implying any unnecessary choice of solution. However, strategic choices may be made a priori, within the classes of solution identified by the business or mission analysis process, such as deciding on a small modular reactor or a Generation IV reactor.
— Place the nuclear facility within its environment and identify any outside entity that interacts with it, or has an influence or expectations. This could include other technical systems (e.g. the electrical grid), the physical environment (providing ambient or seismic conditions, for example), the stakeholders already identified and additional human actors or organizations (e.g. remote support teams or malicious attackers).
— Clearly state the assumptions regarding the environment. This will allow them to be verified and, if necessary, challenged. Assumptions are as essential to good systems engineering as requirements: incorrect assumptions could lead to inadequate system requirements, and assumptions on technical systems or human actors and organizations need to be considered as requirements.

3.2.3.3. Information specific to nuclear facility instrumentation and control systems

The I&C environment is determined by the plant environment (e.g. operators, malicious attackers or the physical environment), the plant architecture and the plant systems. The situations that need to be considered include those determined by analyses such as hazards and risk analysis or vulnerability analysis, which can be made at plant level, plant systems level and I&C systems level.

System requirements for the overall I&C architecture are mainly determined by issues such as safety, computer security, dependability, human factors, cost, procurement, integration and long term maintenance, taking account of facility level concepts such as defence in depth or operations. For safety and computer security, they are often derived from national regulations and international standards and guidelines such as:

— IAEA Safety Standards Series No. SSG-39 [5] (I&C design safety);
— IAEA Nuclear Security Series No. 17-T (Rev. 1) [9] (computer security);
— IAEA Nuclear Security Series No. 33-T [10] (computer security);
— IAEA Safety Standards Series No. SSG-51 [21] (HSI aspects of I&C);
— IAEA Nuclear Energy Series No. NR-T-2.12 [22] (I&C aspects of HFE);
— IEC 60964 [23] (control rooms);
— IEC 61513 [7] (safety);
— IEC 62645 [18] (computer security);
— IEC 62859 [24] (coordination of safety and computer security).

These may have an impact or be impacted by constraints on other disciplines, such as plant layout (e.g. to support the single failure criterion, support the independence of I&C levels of defence in depth and of security zones, and meet requirements for control rooms) and support systems (HVAC and power supplies).

System requirements for individual I&C systems are also strongly influenced by the same issues, but also by the overall I&C architecture. They also need to address functional, performance and operational aspects. Section 4.4.4 and the Appendix provide additional information and examples of the process (see also ISO/IEC/IEEE 29148 [25]).

3.2.4. Architecture definition process

3.2.4.1. Definition and general information

The architecture definition process identifies possible architectural solutions consistent with system requirements and selects the one that best meets the objectives set in the business or mission analysis process.

3.2.4.2. Information specific to nuclear facilities

In addition to the specified system requirements, the overall architecture of a nuclear facility is in large part influenced by safety, security, dependability and cost effectiveness constraints. IAEA Safety Standards Series No. SSR-2/1 (Rev. 1) [8] and a report issued by the Western European Nuclear Regulators' Association (WENRA), Safety of New NPP Designs [26], give requirements and provide guidance for safety. However, as the plant architecture has a profound impact on its I&C system and as modern I&C system costs often comprise a significant part of overall costs, plant architects would be well advised to also coordinate closely with I&C architects.

The overall architecture of the nuclear facility defines important inputs for I&C design, such as the plant structure in terms of plant systems (many of which might need I&C systems), the facility layout, defence in depth, operations, etc. Later, the architecture and design of plant systems will act as the necessary design bases for I&C systems, such as the list and characteristics of I&C functions and sensors and actuators, and the overall architecture of support systems for I&C (HVAC and power supplies).

3.2.4.3. Information specific to nuclear facility instrumentation and control systems

The architecture definition process for nuclear facility I&C systems can be divided into the following processes:

— An architecture definition process for the overall I&C architecture;
— An architecture definition process for each I&C system.

The following standards stipulate the requirements and provide guidance for the safety and computer security aspects of the overall I&C architecture and the I&C systems architecture:

— IAEA Safety Standards Series No. SSG-39 [5] (safety);
— IAEA Nuclear Security Series No. 17-T (Rev. 1) [9] (computer security);
— IAEA Nuclear Security Series No. 33-T [10] (computer security);
— IAEA Nuclear Energy Series No. NP-T-2.11 [27] (I&C architectures);
— IAEA Safety Standards Series No. SSG-30 [28] (safety classification);
— IEC 60987 [13] (hardware aspects);
— IEC 61226 [4] (safety classification);
— IEC 61513 [7] (safety);
— IEC 62340 [29] (common cause failure);
— IEC 62645 [18] (computer security);
— IEC 62671 [30] (digital devices of limited functionality);
— IEC 62859 [24] (coordination of safety and computer security).

The Appendix provides additional information and examples of the architecture definition process.
Overall I&C architecture definition process. IEC 61513 [7] defines the I&C architecture as the organizational structure of I&C systems important for the safety of the facility. In this publication, the overall I&C architecture takes account of all I&C systems, including those that are not important to

safety but are important to plant operation or facility disposal. IAEA Nuclear Energy Series NP-T-2.11, Approaches for Overall Instrumentation and Control Architectures of Nuclear Power Plants [27], provides guidance on issues to consider when developing the overall NPP I&C architecture.

It is a good practice to initiate the overall I&C architecture definition process early in the facility's life cycle, based on assumptions and estimates (e.g. regarding characteristics of the I&C functions to be implemented or available instrumentation) that are afterward gradually consolidated and refined as facility design progresses. Several assumption and estimate scenarios can be considered which may result in different I&C architectures informing plant architects of the impacts on I&C systems of decisions made by people from other disciplines.

Initially, I&C architecture is based on overall facility level concepts such as defence in depth or operations (which, for example, determines what will ideally be manual and automatic), and focuses on organizational and structural aspects such as I&C levels of defence in depth and security zones, main I&C systems and field equipment and their safety class, degree of security, technology and diversity, and data communications. Later, as I&C functions are progressively identified and specified, it allocates them to the I&C systems.

I&C system architecture definition process. IEC 61513 [7] defines I&C system architecture as the organizational structure of that I&C system. It is determined on the basis of the overall I&C architecture and addresses issues such as redundancy, diversity and separation (for fault tolerance), internal and external data communications, selection of I&C platforms and/or main predeveloped components, sizing and computer security (in the case of digital I&C). Thus, there needs to also be interaction with people in other engineering disciplines. These areas include:

— Plant process engineering for detailed specification of I&C signals and required I&C functions;
— HFE for detailed HSI characteristics and 'look and feel';
— Hazard and risk analysis, which needs to take account of hazards and risks caused by postulated I&C malfunctions;
— Electrical power and HVAC engineering (the support systems for I&C), in particular for information on estimated electrical and HVAC power requirements under different plant conditions;
— Operation;
— Maintenance, including day to day maintenance and long term maintenance (with retrofits, upgrades and replacements);
— Validation;
— Commissioning;
— Detailed computer security;
— Dependability analysis to determine I&C system internal redundancy, segmentation and separation;
— Equipment qualification;
— Licensing.

Overall I&C system architecture definition processes provide requirements and design bases for individual I&C systems as follows:

— I&C functions;
— Interfaces between I&C systems;
— Interfaces to sensors and actuators;
— Preliminary assignment of rooms and cable paths;
— Power supply trains.

Based on these inputs, designers of I&C system can allocate functions to cabinets or modules of I&C systems and develop individual I&C system architecture based on certain platform or predeveloped components. As for the I&C level, each I&C system needs to be analysed and verified whether it satisfies applicable requirements. But the full scope of analysis can be done when the design definition process is completed.

3.2.5. Design definition process

3.2.5.1. Definition and general information

The design definition process refines the outcomes of the architecture definition process and provides the detailed data and information necessary for the implementation process.

3.2.5.2. Information specific to nuclear facilities

After the plant architecture definition process has identified the plant systems and defined their interfaces and interactions, the plant design definition process provides detailed solutions for each of them. This is usually done by joint teams from different disciplines.

As plant systems and their operational processes (i.e. how they operate and are operated, not to be confused with engineering processes) are better characterized, information important for I&C design gradually emerges, such as piping and instrumentation diagrams, measurement parameters, actuators, HSI, control posts and functional, performance and safety I&C requirements. It is important that the plant design definition process takes account of I&C constraints, such as avoiding communication from functions of lower safety importance towards functions of higher safety importance, or between functions belonging to different levels of defence in depth.

3.2.5.3. Information specific to nuclear facility instrumentation and control systems

The I&C design consists of the designs of the different I&C systems that make up the overall I&C architecture. The following standards give requirements and provide guidance for facility I&C architecture design:

— IAEA Safety Standards Series No. SSG-39 [5] (I&C design safety);
— IAEA Safety Standards Series No. SSR-2/1 (Rev. 1) [8] (proven engineering practices and design rules);
— IAEA Nuclear Security Series No. 17-T (Rev. 1) [9] (computer security);
— IAEA Nuclear Security Series No. 33-T [10] (computer security);
— IAEA Nuclear Energy Series No. NP-T-2.11 [27] (I&C architectures);
— IEC 61513 [7] (safety);
— IEC 62340 [29] (common cause failure);
— IEC 62645 [18] (computer security);
— IEC 62859 [24] (coordination of safety and computer security).

I&C systems are often designed to include predeveloped commercial off the shelf products. Such products need detailed and rigorous assessment. As this is a multidisciplinary activity, and as organizations responsible for multiple facilities and projects often need some level of standardization, this activity is sometimes analysed during the concept phase and performed as part of the acquisition process.

Disciplines other than I&C may contribute to this process:

— HFE provides:
  • Guidelines for the design of HIS.
  • Verification (often using simulators) to make sure that HSI designs are appropriate. Most guidance suggests that verifying compliance with a standard while conducting the validation would involve operator studies using simulators. HFE can also be associated with late stage (summative) validation.

— Plant systems engineering provides:
  - Algorithmic and response time requirements.
  - Inputs and constant parameters.
  - Test cases and expected results.
— Safety analysis provides:
  - Safety classification.
  - Independence and diversity requirements.
  - Critical functions.
  - Important human actions.
— Computer security provides:
  - Security classification.
  - Security requirements.

Example of design definition processes covering interactions with other disciplines can be found in the Appendix.

As many modern I&C systems are computer and software based, one important activity of the I&C design definition process is the specification of the software architecture, which is defined as the organizational structure of the software of a digital I&C system. Software is here understood in a wide sense and includes programming in HDL and parametrization of devices of limited functionality. It is determined based on the I&C system architecture, and is related to other engineering disciplines such as computer security and licensing.

The following standards give requirements and provide guidance for the safety and computer security aspects of software:

— IAEA Safety Standards Series No. SSG-39 [5] (safety);
— IAEA Safety Standards Series No. SSG-51 [21] (HSI aspects of I&C);
— NUREG-0711 (Rev. 3) [31] (HFE);
— IEC 60880 [11] (software aspects);
— IEC 62138 [16] (software aspects);
— IEC 62566 [17] (HDL aspects);
— IEEE 1023 [32] (HFE).

3.2.6. Systems analysis process

3.2.6.1. Definition and general information

Systems analysis is a process involving several individual analysis activities that constitute a basis for establishing and maintaining a knowledge level that is adequate to support decision making activities throughout the life cycle. The systems analysis process uses information available prior to making key system related decisions during any part of the system life cycle. This information is intended to raise the knowledge level of all decision makers to a level that will result in informed and sound decisions that will then lead to an efficient achievement of system objectives. Because systems analysis includes information that may involve multiple disciplines, these analytical methods are consistent with the principles of systems engineering management (see Section 3.1.3 for information on the types of knowledge to be considered in the performance of systems analysis.)

3.2.6.2. Information specific to nuclear facilities

Because nuclear facility designs are highly integrated, it is necessary to coordinate information and individual system requirements between connected systems and subsystems. The systems analysis process is meant to facilitate this by identifying relevant system integration information for inclusion in critical decision making processes. It is further used as a means of identifying complex system interactions as well as hazards that can be associated with them. Once identified, such hazards can then be addressed through other design processes.

One key analysis performed for nuclear facility is the facility specific accident analysis. Though not within the scope of this publication, this type of analysis provides a basis for safety functions performed by I&C systems. Therefore, it is necessary to coordinate information obtained by the accident analysis with the performance of systems analysis activities and with the development of I&C system requirements.

3.2.6.3. *Information specific to nuclear facility instrumentation and control systems*

I&C systems in nuclear facilities constitute a key element of integration and are usually among the most highly integrated of all facility systems. They are also frequently used as conduits of integration and interaction between plant systems. For example, reactor protection systems (RPSs) interface with many plant systems by measuring process parameters such as pressures, temperatures and levels for the purpose of performing safety functions. The RPS in turn interfaces with reactivity control systems through actuation devices to initiate actuation of the required plant safety functions. In this way, the plant systems are interfaced and integrated with reactivity control systems such as the control rod system through the RPS I&C system.

Analyses that may be included in the performance of systems analysis include the following:

— Failure/hazard analysis;
— Risk analysis;
— Requirements traceability analysis;
— Safety analysis;
— Physical security analysis;
— Computer security analysis;
— Reliability/availability analysis;
— Test coverage analysis;
— Diversity and defence in depth analysis.

Section 6.4.6 of ISO/IEC/IEEE 15288 [1] details a process for performing systems analysis activities, which includes identification and retrieval of input information, performance of activities and tasks and development of specified outputs. The outputs of the systems analysis process include identification of additional analysis needs, validation of assumptions, information such as system interdependencies to support decision making and establishment of requirements with traceability to provide assurance that the systems engineering results will be addressed according to applicable processes. As such, these methods are consistent with the principles of systems engineering management.

3.2.7. **Implementation process**

3.2.7.1. *Definition and general information*

This process realizes the system elements based on the detailed data and information provided by the design definition process. Generally, implementation strategies are established first and include implementation techniques, constraints, risk and countermeasures. The enabling systems or services to support implementation also need to be identified and obtained or acquired in a timely manner. The scope of the implementation process may depend on the process model used.

The implemented elements need to be verified against their requirements and design. They will then be added into the integration process to form the completed system.

Many elements in a nuclear facility are important to safety or plant performance. It is thus essential that their implementation processes are placed under rigorous quality control, from the definition of implementation strategies to the verification of implemented elements. In addition, the implementation processes of elements important to safety may need to be part of the regulatory/licensing process.

An important activity is to prepare or complete the detailed operational and maintenance procedures for the system elements and for the plant systems. As in many technical processes, different disciplines may need to be involved, depending on the nature of the elements to be implemented.

3.2.7.3. Information specific to nuclear facility instrumentation and control systems

The design definition process provides detailed data about an I&C system and its elements. Implementation activities are then performed based on the outputs of the design definition process.

I&C system elements mostly consist of hardware and software. Their implementation may be based on predeveloped I&C platforms and development environments. Development environments are essential enabling systems for the implementation of I&C elements and are generally closely associated (or part of) I&C platforms. They typically include hardware configuration tools, software construction tools and software analysis and testing methods.

The following standards stipulate requirements and provide detailed guidance on hardware important to safety as well as on software implementation:

— IAEA Safety Standards Series No. 39 [5] (safety);
— IEC 60880 [11] (software aspects);
— IEC 60987 [13] (hardware aspects);
— IEC 62138 [16] (software aspects);
— IEC 62566 [17] (HDL aspects);
— IEEE 1012 [33] (verification and validation).

Verification and validation processes are described in Sections 3.2.9 and 3.2.11, respectively. Verification of the implemented elements may be assisted by the integration process.

Since the I&C platforms are selected in conjunction with the definition of the overall I&C architecture or of the I&C system design as described in Ref. [5], this process does not include any activities regarding the selection of platforms. However, the characteristics of a platform need to be considered in the implementation strategies. Commercial off the shelf items for use in nuclear facility I&C applications are also assessed in the architecture and design definition processes and acquired through the acquisition process.

3.2.8. Integration process

3.2.8.1. Definition and general information

The integration process assembles the implemented system elements, verifying at each step compliance with the provisions specified by the architecture and the design definition processes. These provisions need to be fully documented, and include interfaces and interactions between system elements, the interdependencies between functional and physical system elements and the operational processes. The last step results in a fully operational system that satisfies the specified system requirements, architecture and design.

A well prepared system integration strategy and plan is an important input for the project planning process. It also helps mitigate project risks by requiring a systematic, fully documented process for system configuration management and control.

3.2.8.2. Nuclear facility specific information

Integration activities for a nuclear facility bring together first the elements of individual plant systems, and then the plant systems themselves. Coordination is important, particularly when integrating plant systems, which are often engineered by different teams. Also, enabling means, methods and tools often need to be planned and secured early in the life cycle.

For example, not all integration activities can be performed at the facility site. In some cases, it may be preferable or necessary to perform some of them in the factory (where particular system elements or plant systems are implemented) or at specific integration sites so that they can be performed early in the life cycle or they can benefit from adequate tools and expertise.

Due to safety or practical constraints, not all integration tests are possible. In these instances, it needs to be determined how they can be substituted with other forms of verification or justification.

3.2.8.3. Information specific to nuclear facility instrumentation and control systems

Integration activities are frequently required by commonly used IEC standards. Nuclear facility I&C integration typically consists of the following five phases:

(1) Software integration, where the software or logic elements of an I&C system are assembled, and where their interactions are verified using software test tools and equipment and formal software verification tools. As digital I&C systems are often distributed and composed of multiple computing units, software integration may itself be split into two subphases: separate software integration for each individual computing unit, and software integration for the complete I&C system. This phase is performed in the factory.

(2) Software–hardware integration, where the software of an I&C system is integrated with, and tested on, the actual I&C system hardware. Here also, integration may be split into two subphases: separate software–hardware integration for each individual computing unit, and software–hardware integration for the complete I&C system. Specific methods and equipment may be necessary to provide inputs and to collect and analyse outputs, and to perform regression tests. Performance of regression testing is considered at all levels of testing during the entire system life cycle. In some cases, only a simplified I&C system architecture needs to be integrated, provided that there is an adequate justification (e.g. in the case of identical, redundant and independent channels). This phase is also generally performed in the factory. It is followed by factory acceptance tests that validate the functionality of the individual I&C systems.

(3) I&C systems integration, where the I&C systems of the overall I&C architecture are assembled. Here again, specific methods and equipment may be necessary.

(4) I&C and individual plant systems integration, where a plant system is integrated with its I&C, is considered to be part of the integration process of the plant system.

(5) Integration of I&C and plant processes is performed in the framework of the commissioning of the facility. It includes site acceptance tests, which validate the physical and functional integrity of the installed I&C systems.

In cases where there is an HSI, some degree of testing by operators is performed such as an integrated system validation test.

3.2.9. Verification process

3.2.9.1. Definition and general information

The outputs of each engineering process need to be verified against its inputs and the requirements set by the other processes. This activity is considered as a part of that process. The purpose of the verification process is to provide objective evidence that the system, or a system element, complies with its specified requirements. It is different from the validation process, which aims at providing objective evidence that the system satisfies the needs of its stakeholders.

As it is preferable to detect any deviations as early as possible, the verification process is better conducted in step with the architecture and design definitions, as well as the implementation and integration processes.

3.2.9.2. *Information specific to nuclear facilities*

In the case of nuclear facilities, the verification process aims not only at detecting deviations from specified needs and requirements, but also looks for the deficiencies that caused these deviations and tries to provide relevant information for their correction.

It is a good practice to ensure traceability between a system element, its requirements, the verification activities to be performed, the verification activities performed, the raw verification data obtained, and the conclusions drawn. This is particularly useful for regression testing.

In the case of plant systems or system elements important to safety, verification may need to be performed by people and organizations independent of those involved in their design and implementation. Different aspects of independence may be considered, e.g. technical, managerial and financial independence. The degree of independence necessary depends on the importance to safety of the given system or system element.

3.2.9.3. *Information specific to nuclear facility instrumentation and control systems*

The typical relationship between I&C development and verification activities is illustrated in Fig. 2 of IAEA Safety Standards Series No. SSG-39 [5]. The following standards stipulate requirements and provide detailed guidance on hardware important to safety as well as on software verification:

— IAEA Safety Standards Series No. SSG-39 [5] (safety);
— IAEA-TECDOC-384 [34] (verification and validation);
— IEC 60880 [11] (software aspects);
— IEC 60987 [13] (hardware aspects);
— IEC 61513 [7] (safety);
— IEC 62138 [16] (software aspects);
— IEC 62566 [17] (HDL aspects);
— IEEE 1012 [33] (verification and validation).

System verification is based on a variety of techniques:

— Review of the system requirements against stakeholder needs.
— Basic and detailed design review against the system requirements, including both hardware and software aspects.
— Verification of intermediate products that are developed during life cycle, e.g. design verification.
— Software reviews (where correct application of software implementation processes are checked), code inspections (where software source code is examined by persons different from the ones who wrote it) and walkthroughs (where software designers and implementers present and explain their work to persons who did not participate).
— Testing includes structural tests (which aim at covering the logic and structure of software elements) and functional tests (which check the functionality of software elements and integrated software). Unit tests, integration tests, system tests and acceptance tests also use dynamic testing (some of these tests are part of the validation process, and are discussed in Section 3.2.11). Certain test cases for the dynamic analysis may come from other disciplines, which include the organization providing the safety algorithms such as safety analysis or process engineering organizations.

— Tool based static analysis examines software source code without executing it. Static analysis methods include the computation of code metrics, the checking of compliance with coding rules and formal verification methods (see Section 4.3.2).

— In addition to testing with actual hardware, workstation based simulation is a commonly used verification technique for digital technology based designs. It determines their behaviour at various levels of detail and accuracy and at various stages (see IEC 62566 [17] and Section 3.2 of IAEA Nuclear Energy Series No. NP-T-3.17 [35] for an introduction to verification in HDL based designs.

3.2.10. Transition process

3.2.10.1. Definition and general information

The transition process makes the shift between the system development processes addressed in Sections 3.2.1–3.2.9 and the system operational processes addressed in Sections 3.2.11–3.2.14. Its main objective is to install a fully verified, functional and operable system in its environment, together with its enabling systems and trained personnel.

In general, the framework of the transition process includes the activities and tasks shown in Table 1.

3.2.10.2. Information specific to nuclear facilities

The transition process for a nuclear facility involves specific activities and disciplines such as site preparation, transportation and logistics, construction and installation, interfacing with the system environment (e.g. the power grid), training and commissioning. It also involves stakeholders such as the facility general designer, plant systems specialists, suppliers of enabling systems, operators and regulators.

TABLE 1. ACTIVITIES AND TASKS IN THE TRANSITION PROCESS (ADAPTED FROM ISO/IEC/IEEE 15288 [1])

Activities	Tasks
Preparing for the transition	— Define a transition strategy — Identify and define any facility or site changes needed — Identify and arrange training of operators, users and other stakeholders necessary for system utilization and support — Identify system constraints from transition to be incorporated in the system requirements, architecture or design — Identify and plan for the necessary enabling systems or services needed to support transition — Obtain or acquire access to the enabling systems or services to be used — Identify and arrange shipping and receiving of system elements and enabling systems
Performing the transition	— Prepare the site of operation in accordance with installation requirements — Deliver the system to the correct location at the correct time for installation — Install the system in its operational location and interface to its environment — Demonstrate proper installation of the system — Provide training of the operators, users and other stakeholders necessary for system utilization and support — Perform activation and checkout of the system
Managing results of transition	— Record transition results and any anomalies encountered — Record operational incidents and problems and track their resolution — Maintain traceability of the transitioned system elements — Provide key information items that have been selected for baselines

As it is a long, complex and arduous endeavour involving many organizations and disciplines, strategic and rigorous planning and constant coordination are necessary for the transition process. It often places significant requirements that need to be identified, specified and addressed early in the nuclear facility's life cycle.

Planning for the transition process identifies the participants in the process, specifies their roles, tasks, inputs, interactions, dependencies and deliverables, and sets the overall schedule. However, even the most careful of plans will need permanent adjustments in the face of the unexpected and the vagaries of project life, and it is a good practice to prepare contingency measures.

When construction is completed and the enabling systems are operable, commissioning demonstrates proper installation and operators' documentation, operability and consistency with enabling systems, based on a planned combination of tests and inspections. This is done in steps, where the plant systems are gradually integrated and test conditions move gradually closer to real operating conditions (e.g. from cold testing to hot testing). Failed tests or inspections need to be recorded, investigated and traced to appropriate corrective measures. Some of these measures may need to be implemented during the transition process; others may be implemented after the system is put into operation.

The transition process may need to take account of country or site specific constraints and activities, for example in the case of dual purpose technologies, or when calibration equipment with test radiation sources is being used. In particular, some countries have legal requirements for transportation. To ensure the integrity of equipment after the completion of factory acceptance tests (see the first column of Table 1 in IAEA Nuclear Energy Series No. NP-T-3.12 [19]) and before its installation on-site, it may be necessary to develop and test custom container and packaging methods.

3.2.10.3. Information specific to nuclear facility instrumentation and control systems

The transition process for I&C generally begins when factory acceptance tests and other off-site tests are successfully completed on the I&C equipment, which is packed in accordance with the shipping method (protection against moisture, damage, overturning and package acceptance.) If the equipment is not installed upon delivery to the site, then it is stored under safe environmental conditions while awaiting installation. After transportation to the site and verification that the preconditions for installing the I&C are met (for example, installation of adequate anti-seismic devices, access control, power supplies, grounding and HVAC), I&C equipment is checked to make sure it has not been damaged or maliciously altered during transportation, the individual I&C systems are assembled, installed and wired, and their software is loaded, and parameters are set. Tests and inspections are performed to verify that the activities are conducted correctly. Design measures (to be considered by the system requirements definition process) and work procedures are defined to minimize the potential for errors.

Wiring is particularly important, both in term of quantity of work and in term of effects on operation. The I&C is often likened to a central nervous system and its wires to nerves: wiring errors could have adverse or subtle effects. It is necessary to verify that all required wires are correctly connected and routed (for example, to support the single failure criterion when it is required). It is also necessary to verify that there are no extraneous wires that could jeopardize the independence of levels of defence in depth or the independence of systems important to safety with respect to systems of less importance. Also, in the case of multiplexed communications, it is necessary to verify the correctness of addressing. Whereas factory acceptance and off-site tests support the major part of I&C testing, wiring to instrumentation, field equipment and control rooms can be done and verified only on-site during the transition process. Therefore, it is important to coordinate these activities through a rigorous planning and scheduling process.

The facility commissioning tests are the first tests where the I&C is fully connected and integrated to the plant process. When they reveal insufficient plant performance levels, inadequate behaviour or inconsistencies with operational procedures, the I&C may need to be modified (even when it is not the direct cause of these inadequacies).

3.2.11. Validation process

3.2.11.1. Definition and general information

The objective of the validation process is to confirm, based on objective evidence, that in its intended environment the system meets the goals set by the business or mission analysis process and the stakeholder requirements specified by the stakeholder needs and requirements definition process. It determines whether the right system was built, as contrasted with the verification process, which determines whether the system was built correctly. It may be applied to a fully completed system, but also to intermediate engineering artefacts and to system elements.

3.2.11.2. Information specific to nuclear facilities

The validation of a nuclear facility needs to be planned far in advance and implemented along with other processes. In particular, it needs to define objective validation criteria for the specified stakeholder requirements and validate the outcomes of the system requirements definition process: it would be inefficient to realize at the last minute, after years of design and construction and huge expenses, that the facility built is not the right one. The final pieces of evidence are provided by facility commissioning tests, during the transition process.

The validation process is normally performed independently from the processes and products that are being validated. Aspects of independence to be considered include financial, organizational, administrative and technical aspects.

3.2.11.3. Information specific to nuclear facility instrumentation and control systems

The following publications and standards stipulate requirements and guidance on the validation of I& C systems:

— IAEA Safety Standards Series No. SSG-39 [5] (safety);
— IAEA Safety Standards Series No. SSG-51 [21] (HSI aspects of I&C);
— IAEA-TECDOC-384 [34] (software verification and validation);
— IEC 60880 [11] (software aspects);
— IEC 60964 [23] (control room and HSI aspects);
— IEC 60987 [13] (hardware aspects);
— IEC 61513 [7] (safety);
— IEC 62138 [16] (software aspects);
— IEC 62566 [17] (HDL aspects);
— IEEE 1012 [33] (verification and validation).

3.2.12. Operation process

3.2.12.1. Definition and general information

The operation process determines how the system is used to deliver its services.

3.2.12.2. Information specific to nuclear facilities

The operation process for a nuclear facility covers the major part of its life cycle. As it needs to address a very wide range of topics, it cannot be given full justice here: only key aspects and interactions with the other processes will be mentioned.

In addition to effective operation activities, the process involves the following actions:

— Definition of the operation strategy.
— Definition of the concept of operation.
— Identification of the operational constraints that might affect stakeholder requirements, system requirements, architecture, design, transition, validation and maintenance (e.g. operating and maintenance staff/operating cycle time/outage time).
— Identification, specification and development or procurement of enabling systems (e.g. training simulators), services and material (e.g. operating procedures and corresponding documentation) needed for operation.
— Personnel training and qualification.
— Collection and analysis of operating experience, and recording, investigating and tracking of incidents and accidents.

The process is an essential source of inputs to the:

— Stakeholder needs and requirements definition process.
— System requirements definition process.
— Architecture definition process (e.g. for control rooms, the ability to perform maintenance during operation or to perform efficient outages).
— Design definition process (e.g. for maintenance, periodic testing, incident and accident management and day to day operation).

3.2.12.3. Information specific to nuclear facility instrumentation and control systems

This process has a strong influence on I&C since a significant part of plant operation occurs through I&C. IAEA Nuclear Energy Series No. NR-T-2.12 [22] provides extensive information and guidance, including the concept of operation, which describes the system of interest (here, a nuclear facility) from the viewpoint of the individuals who will operate it.

3.2.13. Maintenance process

3.2.13.1. Definition and general information

The maintenance process aims at ensuring that the system can provide its services throughout its planned operational lifetime.

3.2.13.2. Information specific to nuclear facilities

The lifetime of a nuclear facility is typically several decades (up to 60, or even 80 years). Thus, it is necessary to take into account not only day to day maintenance, where the facility and its systems are kept in an as-designed state, but also long term maintenance with retrofits (i.e. form, fit and function module replacements), upgrades (replacements of a plant system with limited changes to the rest of the plant) or modernizations (significant changes in multiple plant systems, plant architecture and/or plant performance).

For equipment that is in a poised state most of the time (e.g. many safety systems), periodic testing might be necessary. For equipment that is in an active state most of the time (e.g. many normal operation systems), on-line monitoring could be considered. Periodic testing and on-line monitoring often require specific design and operational measures. As any additional equipment will also need to be maintained, an adequate balance needs to be determined.

A number of IAEA publications address the issue of maintenance:

— Safety Reports Series No. 42 [36] (safety culture);
— IAEA Safety Standards Series No. NS-G-2.6 [37] (maintenance, surveillance and in-service inspection);
— IAEA Nuclear Energy Series No. NP-T-3.8 [38] (maintenance optimization);
— IAEA-TECDOC-960 [39] (regulatory surveillance of safety related maintenance);
— IAEA-TECDOC-1138 [40] (safety related maintenance);
— IAEA-TECDOC-1335 [41] (configuration management);
— IAEA-TECDOC-1383 [42] (optimizing maintenance programmes);
— IAEA-TECDOC-1532 [43] (operation and maintenance);
— IAEA-TECDOC-1590 [44] (reliability centred maintenance);
— IEC TR 62096 [45] (decision on modernization).

3.2.13.3. Information specific to nuclear facility instrumentation and control systems

The same distinction between day to day and long term maintenance is applicable here. It is also important to consider the following:

— *Maintenance of the I&C system.* Maintenance related decisions may have a strong impact on the architecture of an I&C system, for example if it is to be maintained while the plant is in normal operation. IAEA-TECDOC-1402 [46] provides guidance on the subject.
— *Supporting and optimizing maintenance.* I&C systems can support and optimize the maintenance of other plant equipment, e.g. through on-line monitoring or data reconciliation (where models and redundant sources of information are used to detect inconsistent data). However, a balance needs to be found between benefits such as improved diagnostics (i.e. identification of failed components) and prognostics (i.e. prediction of impending failures), and additional costs and design complexities, including the need to maintain additional sensors and to cope with their postulated failure.

3.2.14. Disposal process

3.2.14.1. Definition and general information

The disposal process handles the end of life of the system or of system elements and their disposal, as appropriate.

3.2.14.2. Information specific to nuclear facilities

Disposal of nuclear facility elements may occur all along the plant operational lifetime. The disposal process itself may take many decades. Not only do possibly hazardous materials need disposal, the facility and many of its systems need to be monitored for long periods of time. In addition, specific actions are necessary after the final shutdown of the facility (e.g. allowing cooling of the fuel pool for a certain period, monitoring functions, power supplies, etc.). Constraints on disposal may be expressed early in the nuclear facility's life cycle and contribute to the system requirements definition process.

3.2.14.3. Information specific to nuclear facility instrumentation and control systems

Retirement and ultimate disposal of I&C systems occur when I&C components are replaced or upgraded and when the nuclear facility is removed from service. Also, specific I&C functions and equipment may be needed to support the disposal process.

3.3. TECHNICAL MANAGEMENT PROCESSES

3.3.1. Project planning process

3.3.1.1. Definition and general information

This process develops and coordinates work plans for projects concerning the system. In particular, it sets the scope of project management and technical activities, specifies the inputs, activities, deliverables and achievement criteria of processes, establishes schedules, and identifies the resources necessary to accomplish tasks. It begins ahead of project activities, which it defines in the planning documentation. It is also updated throughout the project with revisions to account for progress and to address any issues.

Project planning can be used at different stages of the system life cycle, including design, testing, operation, maintenance and retirement. The planning process also involves the establishment of a hierarchy of coordinated plans that when executed provide an organized framework for completing required actions in conjunction with all planned activities for the system. As such, the use of one or more high level master plans that integrates the individual processes plans together is often necessary to achieve this. Quality plans are often used for this purpose.

3.3.1.2. Information specific to nuclear facilities

Because of the very large number of stakeholders, organizations, engineering disciplines and teams involved, project planning is an essential and strategic process for nuclear facilities. In particular, it identifies needs for coordination that individual stakeholders, organizations, engineering disciplines and teams might not recognize. Also, as a project often covers only part of the lifetime of a facility, project planning can give it a life cycle perspective by ensuring that the right set of stakeholders and disciplines are involved. Different organizations contributing to the same project may have their own project planning process, but coordination is needed to ensure consistency and efficiency.

The practical outcomes of project planning are expressed in plans. A plan defines the scope of practices and sequences of activities needed for a particular issue. It also identifies the competences, roles, resources, enabling systems and tools required for the activities planned, and defines measures (with additional stakeholders, disciplines, teams and activities), ensuring that they are available when needed. A plan also identifies the input documentation necessary for completing planned activities as well as the output documentation upon completion of these activities. Thus, plans provide important inputs for schedule and resource management.

Plans may be multitiered: generic plans may specify general provisions and make facility level decisions, whereas system level or project level plans may customize generic plans to the specific needs of particular systems or projects. For example, a generic I&C plan may specify general provisions for I&C and a framework for the plans of individual I&C systems according to their safety class. It is a good practice to strictly define the scope and applicability of such plans and to assign activities to well identified project participants.

Examples of plans:

— *Quality assurance plans* set out the quality practices and activities, ensuring high quality and compliance with all specified requirements and constraints. This includes making sure that all teams involved coordinate as necessary and in time, and that resources are available when needed. Quality assurance, or quality plans, are often used to facilitate coordination and interactions among other planning activities.

— *Requirements management plans* set out the practices and activities leading to the specification of requirements (including stakeholder requirements, system requirements, system elements requirements and requirements for particular technical processes). They ensure the active involvement of the stakeholders and those in the relevant disciplines, addressing unrealistic expectations, ensuring that the specified requirements are appropriate for all situations, are unambiguous, achievable, properly identified, recorded, traced and retrievable. The requirements management plan also needs to establish processes for traceability between requirements and design, between requirements and tests, and needs to include activities to confirm that all requirements are met.

— *Integration plans* set out the practices and activities necessary for the integration process. These include identification of interfaces, specification of interface requirements, establishment of interface design characteristics, and testing to ensure that interfaces are correctly implemented.

— *HFE interface plans* are used to address activities necessary to integrate HSI into the system. This plan sets out practices and activities necessary to support the HFE design, such as the concept of operations, HSI style guide, procedure writing and operator testing.

— *Installation plans* set out the practices and activities needed to successfully implement the nuclear facility and its systems on-site, and to perform all necessary on-site verification.

— *Maintenance plans* set out the practices and activities needed to maintain the nuclear facility and its systems throughout operation.

— *Operation plans* set out the practices and activities necessary to the operation process.

— *Security plans* set out the necessary security practices and activities, such as vulnerability analyses, specification of security requirements, and implementation of security measures.

— *Verification and validation plans* set out the practices and activities for the verification of technical processes and their outcomes. They also set out procedures to ensure that errors that are detected are appropriately analysed, reported, corrected and reassessed.

— *Configuration management plans* set out the practices and activities for configuration control and management activities. They make sure that the necessary infrastructure is available when needed and that staff are trained.

The extent to which project planning is applied is usually determined early and often depends on the importance of the issue and on the scope and complexity of the activities to be performed.

3.3.1.3. Information specific to nuclear facility instrumentation and control systems

Information which is relevant at the nuclear facility level also applies, to a large extent, to I&C systems. Some additional issues include:

— *Assessment and selection of off the shelf I&C platforms and products.* Owing to demanding safety and computer security requirements, but also because of the potential impacts on plant reliability, safety and security and to the long operational life required (several decades), such assessments are extensive and arduous, involving multiple organizations, stakeholders and disciplines.

— *Off-site integration and testing.* These are particularly important when the I&C architecture is composed of systems from different vendors (which is often the case). The organization responsible for overall off-site integration and testing (which may be required before installation on-site) and for providing the necessary means needs to be carefully selected.

— *Integration of I&C systems in facility level training and/or HFE simulators.* These simulators are often significant systems on their own. To ensure fidelity with respect to the real facility, it often necessary to make early plans to ensure that I&C systems are correctly represented.

— *Independent verification and validation.* This is required for the licensing of safety digital I&C systems. As it may represent significant effort and time and involve multiple organizations, planning is necessary.

3.3.2. Configuration management process

3.3.2.1. Definition and general information

The purpose of the configuration management process is to manage and control system elements and their configuration over the system life cycle. It also manages consistency between a product and its associated configuration definition within the configuration baseline for an I&C system.

3.3.2.2. Information specific to nuclear facilities

Effective configuration management is essential for efficient, safe and secure day to day operation and maintenance of the nuclear facility and its systems, but also for future improvement and renovation projects. Due to inadequate configuration management early in their life cycle, some facilities have needed to spend a great deal of effort to reconstitute their design basis.

IAEA publication, Configuration Management in Nuclear Power Plants (IAEA-TECDOC-1335) [41], provides concepts for configuration management, based on operating experience, at the nuclear power plant level. The concept of product or plant life management is an important aspect of facility configuration management and is discussed in elsewhere in this publication.

The need for effective configuration management is recognized in other IAEA publications [8, 47] that provide guidance on its application to nuclear facilities. Another IAEA publication focuses on information technology to support the configuration management process [48].

3.3.2.3. Information specific to nuclear facility instrumentation and control systems

I&C life cycles are normally iterative in nature and therefore the management of changes is a key element of configuration management for I&C in terms of establishing a baseline. Section 2.6 discusses the use of iteration and recursion in engineering processes, which needs to include the need to retain a configuration baseline. The following publications and standards provide requirements and guidance on configuration management for I&C systems:

— IAEA Safety Standards Series No. SSG-39 [5] (safety);
— IEC 61513 [7] (safety);
— IEEE 828 [49] (software configuration management).

The following publications and standards stipulate requirements and guidance on configuration management for I&C system software:

— Safety Reports Series No. 65 [47] (nuclear facilities);
— IAEA-TECDOC-1651 [48] (configuration management);
— IEC 60880 [11] (software aspects);
— IEC 62138 [16] (software aspects);
— IEC 62566 [17] (HDL aspects).

As part of an effective configuration management process, items that have to be included in the arrangements (commonly referred to as configuration items) are I&C systems and equipment important to safety and their configuration (both physical and other settings), operational and safety case requirements, limits and conditions, maintenance requirements and instructions. These arrangements need to cover the life cycle of I&C systems, from requirements capture through design, manufacture, installation, commissioning, operation, maintenance, modification and decommissioning.

The elements of configuration management at the organizational level for I&C systems have to include:

— Effective planning;
— Identification and management of change control;
— Asset management arrangements (also referred to as status accounting);
— Regular review to ensure that the configuration is being maintained.

In addition, there needs to be suitable oversight of the supply chain and contractors such that potential changes (e.g. arising from modifications to I&C equipment, updates to system or equipment software) that may adversely impact the configuration baseline can be identified and managed. This may form part of the intelligent customer role that needs to be applied proportionately to, as a minimum, ensure verification of the configuration of items important to safety.

3.3.3. Information management process

3.3.3.1. Definition and general information

The purpose of the information management process is to generate, obtain, confirm, transform, retain, secure and dispose of relevant, and possibly confidential, engineering information (such as technical, project, organizational, agreement, user information, operational data, failure data), and to disseminate it to designated parties in a timely manner during the system's lifetime, and possibly beyond if necessary. It is thus an essential means of coordination between the stakeholders, engineering disciplines and teams involved in a project or concerned with the system. Information models may be used to make sure that the information to be managed suits the needs of the parties concerned. Such models are not determined by the information management process alone: on the contrary, all the other engineering processes need to be used. For example, an information model could specify:

— The various types of information to be obtained and managed.
— The various relationships between types or pieces of information.
— Information expected or produced by the different engineering processes.
— Milestones (as defined by the project planning process) and expected information at each milestone.
— Various features for specific information, such as availability (e.g. planned/expected, draft, or available), status (e.g. validated or not, and by whom), or access control (e.g. who is allowed to gain access).
— Workflows, i.e. how specific information is shared and circulated to reach a final disposition.
— Dissemination, i.e. who needs to be alerted in the various engineering processes regarding availability or changes.

Product life cycle management systems are an emerging category of software tools that can be used to support the management of information.

3.3.3.2. Information specific to nuclear facilities

Multiple organizations usually need to work together in the engineering of a nuclear facility: owner organizations, design organizations, equipment suppliers, licensors, operating organizations, etc. Each is likely to have its own information management processes, models, tools and databases. Thus, one of the goals of the information management process, and of the agreement processes, is to ensure that all of them are interfaced and interoperable as necessary to serve the needs of the nuclear facility's engineering, taking account of computer security and intellectual property constraints.

Also, a single information model addressing all the engineering needs of nuclear facility would be extremely complex, and agreement between all the parties concerned would be very difficult to obtain. It is preferable to have separate, more manageable, information models and databases, each focused on specific issues and interconnected as necessary to the other models.

Lastly, account needs to be taken of the fact that nuclear facilities have very long life cycles compared with computer based enabling tools. Provision needs to be made to ensure that information can be transferred from one tool version to the next, or from one obsolete tool to a more up to date one.

3.3.3.3. *Information specific to nuclear facility instrumentation and control systems*

As the central nervous system of the plant, I&C is particularly concerned with information management, and I&C specialists need to contribute to the information management process. In addition to facility level information models, I&C specific information models could be used for:

— The I&C configuration baseline;
— The overall I&C architecture;
— The architecture of individual I&C systems;
— The I&C functions, as required by other engineering disciplines and teams.

3.4. REGULATORY/LICENSING PROCESS

3.4.1. Information specific to nuclear facilities

The adoption of systems engineering methods is a means of establishing the basis for safety and security conclusions to support determinations of regulatory compliance. The processes adopted by various national regulatory bodies can vary between Member States and, as such, it is important that licensees understand the regulatory requirements applicable to each nuclear facility development. As stakeholders in systems engineering processes, the primary interests of regulators are system safety and security. Licensing processes involve licensee provided justification(s) for claims that requirements for safety and security at a nuclear facility can be satisfied by the practical implementation of structures, systems and components (SSCs). These justifications are evaluated to determine if the regulatory requirements are being met. The evaluations can involve performance of a series of activities in relation to the licensee claims and justifications. These claims and justifications are supported by evidence to enable one or more safety and security conclusions. This evidence is used to provide a traceable basis for each of the conclusions reached.

An example of an evaluation process is the following:

— Regulatory analysis to determine applicable regulations and regulatory guidance. These are used as criteria for making subsequent determinations against the claims for safety and security being made for the SSCs, within the context of the nuclear facility's design and operation.
— Evaluation of the proposed justification by reviewing evidence provided by the licensee to support safety and security claims.
— Determination if the regulatory requirements identified during the regulatory analysis activity are satisfied.
— Collection of comments and requirements from regulators on any gaps in the justification or evidence provided as part of a systematic review phase, and organize a structured process to address these issues.
— Documentation of conclusions and provision of a traceable basis for each conclusion.
— Summary of the evaluation's conclusions and determination of acceptability for the proposed activity.

From the licensee's perspective, the regulatory process often involves the development of regulatory compliance claims and collection and/or production of evidence justifying that the regulatory requirements for nuclear safety and security that are applicable to the claims made for a nuclear facility are being met. To that end, the licensee may need support from other organizations that contributed to the design and implementation. Systems engineering processes provide a cross-disciplinary approach for the development of systems, which also establishes an effective means of collecting the evidence needed to support regulatory determinations. This needs to be taken into account during the licensing process.

Many regulatory requirements are not prescriptive and leave room for different approaches and solutions. Therefore, it is helpful to plan the regulatory/licensing process by including consideration of regulator needs to support early engagement between the licensee and regulator on the adequacy of contemplated approaches and principles that are likely to underpin possible solutions. This early engagement between the licensee and regulator is usually performed in advance of the completion of the design and its implementation, after which subsequent changes can have considerable impacts on costs and can cause delays.

The licensee typically provides a series of documents on the nuclear facility to the regulatory authority to support its assessment of the safety case and other related information (fault analysis and engineering schedules, operating rules and instructions, etc.) to determine whether this satisfies regulatory criteria, referring to systems engineering outputs as appropriate.

3.4.2. Information specific to nuclear facility instrumentation and control systems

The regulatory/licensing process is applied in broadly the same manner to the overall I&C architecture and to the individual I&C systems. For an individual I&C system, the information that constitutes the basis for conclusions can consist of one or more of the following:

— Traceable and verified system requirements;
— Details of the I&C system design and its implementation;
— Configuration baseline;
— Test and commissioning results;
— Verification of system performance or activity characteristics, including its interaction with other I&C systems and equipment at the facility;
— Analysis of processes used to complete the licensing activity;
— Confirmatory review of documentation associated with process execution.

The engineering processes described in the standards referenced in this publication are recognized and endorsed by many regulators for establishing conformance to nuclear safety and security requirements. Regulators also recognize the benefits that systems engineering processes provide in addressing complicated system interactions.

4. SUPPORTING METHODOLOGIES

To evaluate a solution against applicable requirements, systems engineering deals with large amounts of information, from contract clauses, requirements and task schedules to specific and detailed data about solutions and their constituents. To support the engineering of safe and state of the art systems, rigorous methodologies are needed.

This section provides a description of frameworks and methodologies that can be used to support systems engineering processes. It is neither exhaustive nor prescriptive: other frameworks and methodologies can be considered, and each organization needs to make its selection of methodologies, frameworks and tools based on the needs of each project.

4.1. REQUIREMENTS ENGINEERING

Requirements engineering is a discipline concerned with discovering, developing, analysing, verifying, validating, communicating, documenting, managing and maintaining requirements. It involves several systems engineering processes, as discussed in Section 4.1.3.

Stakeholder and system requirements are particularly important, since they are the basis of architecture definition, design, integration, verification and validation. Thus, it is essential to initiate the corresponding processes early in a project, with requirements engineering as a core discipline facilitating not only the assembly of requirements but also the implementation of traceability throughout the project. It also helps reduce the likelihood of changes late in the system's life cycle, which can have significant adverse impacts on cost and schedule.

Requirements engineering is not concerned only with stakeholder and system requirements and needs to be applied recursively. Indeed, stakeholder and system requirements are identified first, but as concepts are developed and the system is designed, these initial requirements evolve into requirements allocated to subsystems, components and design elements (e.g. software). Requirements engineering is thus also at the core of the architecture and the design definition processes.

For an I&C system, to take account of the full scope of inputs, it is preferable to frame its requirements in the context of the nuclear facility (or at least of the plant systems concerned) and of the overall I&C architecture.

4.1.1. Guidelines for definition of requirements

When defining requirements, the following guidelines, based on the lessons learned across different industries, may be considered:

— Requirements have to consider the items to which they apply as 'black boxes' and express what needs to be achieved, not how to achieve it. It is best not to imply support for any particular solution, product or technology. Doing so would obscure the real purposes of the item and complicate the verification of the requirements. It would also restrict the scope of possible solutions and thus the potential for optimization.
— Thus, requirements need to be as complete as possible and express all necessary capabilities, characteristics, constraints and quality factors. The Cranbrook Manoeuvre, described in Section 4.4.4, illustrates the need to take account of the interfaces and interactions of the item with its environment and the different situations (normal and abnormal) it may face over the system's life cycle.
— Each defined requirement needs to be checked that it is necessary and contributes positively to the fulfilment of stakeholder requirements. Overambition (particularly as digital I&C technologies offer virtually unlimited functional capabilities) could lead to unnecessary complexity and costs, both in design and operation.
— Requirements need to be expressed as precisely as possible, in a clear and concise manner, so that those concerned do not interpret them differently, and so there are objective verification criteria for deciding whether they have been satisfied (e.g. through test, analysis or inspection). For example, statements such as 'easy to use' or 'ideally' are guaranteed to create confusion. Several templates are proposed in ISO/IEC/IEEE 29148 [25].
— Requirements need to be feasible. Before they are finalized, it is necessary to ensure that there are affordable solutions for problematic requirements in terms of cost and schedule, and in terms of design, licensing and operation.

— Requirements from applicable codes, standards (e.g. SSR-2/1 (Rev. 1) [8]) and regulations need to be identified and addressed.

4.1.2. Requirements attributes

Various attributes may be used as appropriate and applicable to characterize a requirement, including:

— *Identification*: Unique name or number.
— *Targets*: Systems, subsystems, components or elements to which the requirement applies.
— *Specification*: In text and/or diagrams, possibly in multiple languages, including natural language and machine interpretable formats.
— *Version*: Unique indicator of the requirement version.
— *Rationale*: Explanation of why the requirement is needed.
— *Sources*: Traceability links to the upstream elements that led to the requirement.
— *Potential solutions*: Indication of possible ways of implementing the requirement if its feasibility is in doubt.
— *Verification method*: How compliance to the requirement is to be verified.

Some attributes may be used provisionally during the requirements definition processes when a requirement is not finalized:

— *Status*: Indicator of the degree of finalization of the requirement.
— *Risk*: Indicator of the likelihood that the requirement cannot be satisfied or will not be retained.
— *Urgency*: Indicator of the urgency of the requirement.
— *Alternatives*: If one requirement has to be chosen among several. The priority attribute may then indicate the preferred one.

Other attributes for nuclear facility I&C functional requirements include:

— Actuation priority;
— Safety category;
— Level of defence in depth;
— Inputs;
— Outputs;
— Hazards.

For example, a requirement for reactor trip in case of high neutron flux has its origin from a safety analysis requirement, has priority over reactor power control functions, and poses a risk for spurious actuation.

4.1.3. Linkage with systems engineering processes

Requirements engineering is part of, and interacts with, many engineering processes:

— *Stakeholder needs and requirements definition process.* Develops stakeholder requirements, which are the basis for elaborating system requirement.
— *System requirements definition process.* Develops system requirements, which provide a technical, practical and verifiable view of a system constituting an operable solution to stakeholder requirements. Within this process, requirements analysis can be used to provide a balanced set of requirements to ensure the integrity and validity of system requirements.

— *Architecture and design definition process.* These provide an architectural and detailed description of the system, its constituents, their behaviours and their interfaces and interactions which together satisfy the system requirements. Through this process, the requirements of plant systems and of the overall I&C systems are allocated to their constituent parts. The requirements of I&C systems are then allocated to their own subsystems. Traceability between system requirements and architecture and design is a good practice.

— *Systems analysis process.* Identifies issues that need to be addressed in the requirements, in particular in I&C requirements.

— *Regulatory/licensing process.* Places stringent requirements on systems important to safety or security.

— *Integration, verification, transition, validation, operation, maintenance* and *disposal processes.* If the corresponding experts were not considered as stakeholders from the beginning, they may place constraints that need to be addressed in the requirements, in particular in the I&C requirements. However, that would not be a best practice since such additional requirements would be imposed late in the life cycle, with possible serious impacts on costs and schedules.

— *Verification process.* This relates to the outcomes of the stakeholder needs and requirements and system requirements definition processes, to ensure that the specified requirements have the attributes listed in Section 4.1.1 and that they constitute an adequate answer to the stakeholder requirements. Also, as the verification plan is based on system requirements, a method of verification needs to be specified for each requirement, which then needs to be checked that it is unambiguous, complete, traceable, verifiable and consistent with other requirements.

— *Validation process.* Provides evidence that the system complies with stakeholder requirements, achieving its intended use in its intended operational environment.

— *Configuration management process.* Ensures that requirement baselines and changes are properly identified, recorded, verified, approved and released in a structured and controlled way.

— *Information management process.* Ensures that the necessary information regarding requirements is communicated in a timely manner and in an appropriate format to all concerned.

4.2. ASPECTS OF FUNCTIONAL REQUIREMENTS ENGINEERING FOR INSTRUMENTATION AND CONTROL

4.2.1. Definition of functional requirements

The functional requirements are defined in a specification which can be considered as part of the system requirements specification, or it can be a separate document. The scope of the functional specification process is to identify and define all I&C functions and their safety categorization.

The I&C function specification process is an iterative process in which several disciplines are involved. It is recommended that the following attributes of the process be defined at the beginning:

— Involved disciplines;
— Their responsibilities and output (data/documents) to be provided;
— Time sequence and level of detail (what will be delivered and when);
— Identification of interfaces;
— List of I&C function attributes and criteria for assignment.

IEC 61226 [4] provides generic criteria for the identification and categorization of all functions necessary to fulfil the main safety functions in all plant states. Functions are defined and categorized regardless of the physical means that are implemented to fulfil them. In the framework of IEC 61226 [4], the functions to be categorized are performed by I&C systems. Accordingly, they are called I&C functions.

Safety functions need to be assigned to different categories according to their safety significance. The category determines the class to be assigned to the I&C systems that perform these functions and the class to be assigned to the electrical power systems that support the functions (including those that support the operation of the I&C equipment). Then the class determines the design and quality requirements for these systems.

IAEA Safety Standards Series No. SSG-30 [28] gives guidance on the categorization of functions and classification of SSCs according to their safety significance. According IEC 61513 [7], the identification of all safety related functions needs to be done at an early stage in the design of nuclear facilities. As it may not be possible to identify in detail all functions at an early stage in the design process, it is possible to continue the process of identification and categorization of the safety and non-safety related functions iteratively throughout the design stage. It is also advisable to structure the I&C safety functional requirements specification process in stages so that process engineers can deliver important inputs to the I&C engineers and parallel work is possible.

During the I&C functional specification process, the following items are identified as central issues:

— Common understanding/harmonization concerning I&C design aspects (e.g. I&C related fault postulates, separation and diversity requirements, spurious actuations, etc.), which need to be considered by process engineers during the elaboration of functional requirements before providing the necessary inputs to I&C discipline.
— Codes and standards applied.
— Identification of postulated initiating events and their frequency of occurrence (including I&C related fault postulates).
— Specification of I&C functions and related parameters.
— Severity of consequences if I&C functions fail.
— Plant state to be reached (safe state/controlled state) in the case of failure of control functions.
— Plant operating state (operation, cold shutdown, etc.).
— Application of the defence in depth and diversity concept in relation to I&C functions.
— Safety categorization of the I&C function.

I&C functions can be defined at several levels, since more precise definitions are possible when moving from level to level. The example of safety functions in Fig. 5 has four levels.

Level 1 specifies the safety functions derived from the fundamental safety objectives. Level 2 details the process functions needed to ensure safety functions, including the safety categorization and defence in depth level. This level may also generate other process functions for other reasons such as asset protection. Level 3 defines the I&C functions needed to perform process functions. I&C functions are described in the form of logical diagrams with interfaces to measured parameters, actuators and other functions. It provides the necessary basis to allocate functions to individual I&C systems and identify the interfaces between them. Also, level 3 functions are used as input for elaboration of more detailed functional requirements on the next level. The level 3 functional specification typically contains the following information:

— Input signals (measurement data);
— Logic operations/function definition;
— Type of function (control, limitation, reactor trip, etc.);
— Response time target values;
— Redundancy;
— HSI requirements;
— Failure annunciation and safe state for I&C functions;
— Accuracy requirements;
— Alarm/measurement shown in the main control room/emergency control room;
— Request for diversity (only for defence in depth level 3).

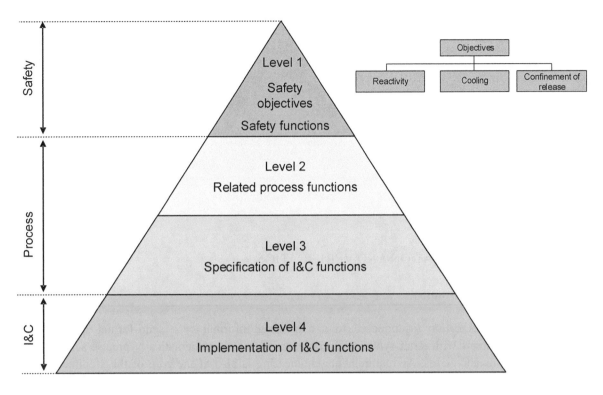

FIG. 5. Levels of safety functions.

Level 4 refines I&C functions (including I&C related details like signal conditioning and filters for the inputs, test inputs to the system, diagnostics functions, etc.) allocated to the I&C system. The first three levels are defined by process engineers. The fourth level is defined by I&C engineers based on inputs from process engineers. and is implemented as part of the architecture definition process. This process can be adapted for non-safety related systems.

4.2.2. Early validation of functional requirements

Late functional design changes lead to comprehensive modifications of scope (affecting large scope of domains) and can endanger the overall project cost and time schedule. The reasons for late functional design changes are incomplete functional requirements or undetected design faults in the implementation of functional requirements. Therefore, a systematic validation of the functional requirements (e.g. the level 3 functional specification presented in Section 4.2.1) needs to be performed before they are finally released.

The activities for the functional requirements validation process depend on the complexity of the requirements. It is good practice to consider the use of simulation tools or simulators in case of complex requirements for the validation and to involve shift operators at an early stage in checking functional requirements related to plant operation.

Using these tools, test vectors can be generated automatically based on test scenarios. Then, simulations with the designed solution (functional requirements) can be executed and the behaviour of the specified system can be observed. Any violation of requirements by the candidate solution can be detected. With such an approach, ambiguous, incorrect, missing or conflicting requirements can be detected.

4.2.3. Interlinkage with systems engineering processes

Functional requirements engineering interacts with the same systems engineering processes as requirement engineering. They are:

— Stakeholder needs and requirements definition process;
— System requirements definition;
— Architecture and design definition process;
— Verification;
— Validation;
— Configuration management.

4.3. FORMAL SPECIFICATION AND VERIFICATION

4.3.1. Specification formality

A formal specification (as opposed to a non-formal/informal or a semi-formal specification) describes requirements with strict syntax and grammar rules and mathematically precise semantics to ensure they are unambiguous. As requirements engineering is also at the core of the architecture and the design definition processes (see Section 4.1), a formal specification may also describe a solution to requirements. Formal specifications are also called formal models.

A formal specification is typically written in the appropriate language. The formal nature of the language is based on mathematics and uses concepts from algebra, logic and set theory. Non-formal/informal specification languages (such as natural language) have flexible construction rules and use a large number of language elements, some of which have multiple meanings, thus leaving room for different interpretations and ambiguity. Semi-formal specification languages mix formal and non-formal features.

Of particular importance to the use of formal methods and descriptions is the fact that requirements specification and conceptual design are critical activities for project success because the correction of undetected errors in these artefacts that are found late in the development life cycle induces exponentially increasing costs and delays in projects.

The advantage of formal specification languages is that, beyond unambiguity, they provide rigorous and systematic conceptual frameworks which, within their scope of application, may guide and organize the thought process and limit the likelihood of errors. They may also benefit from tool support which may help reveal contradictions among requirements (such that no solution can satisfy them all) or between a solution and its applicable requirements. The disadvantages are that current formal methods need specific expertise, their scope is typically limited and they require extra effort initially (but these extra resources can be reclaimed in the later project stages). Also, one needs to acknowledge that a model is not the actual product: its accuracy may be limited, and it may rely on assumptions that need to be stated and justified.

The exchange of information by experts in different engineering disciplines during facility design can benefit from using formal languages. The largest contributor to systematic errors during design are the specification errors resulting from the incompleteness of the requirements or the misunderstanding of the requirements and constraints prescribed by those in one engineering discipline to the other. The lack of ambiguity of formal languages and the possibility for formal analysis significantly reduce the potential for misunderstandings and incompleteness.

4.3.2. Formal verification

Formal verification is the subdomain of verification in which verification is based on formal models and it is carried out using formal methods. According to Ref. [50]:

"A formal method is a set of tools and notations (with a formal semantics) used to specify unambiguously the requirements of a computer system that supports the proof of properties of that specification and proofs of correctness of an eventual implementation with respect to that specification."

The same definition may be applied to models of non-computer systems. In system development, formal methods are description, modelling and analysis techniques based on mathematical methods that support tools for specifying, developing and verifying system models with clear and mathematically precise syntax and semantics, and provide proven correct analysis procedures for these tasks.

4.3.3. Interlinkage with systems engineering processes

Formal specification is an instance of requirement specification and can be used as a part of the following systems engineering processes:

— Stakeholder needs and requirements definition process;
— System requirement definition;
— Architecture definition process;
— Design definition process;
— Systems analysis process.

Formal verification is an instance of specific activity in the verification process.

4.4. MODELS TO SUPPORT SYSTEMS ENGINEERING

4.4.1. Models

A model is a selective and simplified representation of an actual system, entity, phenomenon or process. Objectives help in communicating and/or understanding aspects of interest. As opposed to physical models based on mechanical, electrical or other concrete elements, digital models (i.e. models that can be exploited using computers) are generally expressed in modelling languages based on a combination of physics, mathematics, logic or procedural notations. Though physical models may play an important role in systems engineering, the following focuses on digital models, and 'model' stands for a 'digital model':

— *Geometric 3-D or 2-D models*. These define volumes, surfaces, shapes, topologies and sometimes movements. For I&C systems, such models may be used, for example, to make sure that there is enough room to install, accommodate, test, operate, maintain and replace all the I&C cabinets, or that cable paths for safety I&C systems adequately support independence.
— *Geographical models*. Describe the topography and the geology of a given area.
— *Engineering databases, holding static characteristics of components and system designs*. For I&C systems, these could, for example, cover the failure rates, the expected lifetimes and the required operating conditions of different types of I&C components and I&C controlled equipment.
— *Requirements and assumptions models*. These models include formal requirements definition, elaboration and traceability up to the related object, documents and test cases. Formal definitions

are those given in languages with well defined syntax and semantics so that there is no ambiguity and they can benefit from extensive tool support, such as simulation or automatic verification of compliance to requirements. For I&C systems, they can help ensure that I&C functional requirements are unambiguous and correctly reflect the intentions of the engineers involved.

— *Functional models*. Define in a hierarchical structure a set of functions that need to be implemented. They help understand how I&C functions contribute to plant or plant system functions and to levels of defence in depth. They also help make sure that the safety categorization of I&C functions is consistent with the categorization of the higher level functions to which they contribute.

— *Probabilistic models*. Used to evaluate achieved level of safety or availability against target values.

— *Economic models*. These cover costs and revenues.

— *Operational procedures*. These specify the human actions to be performed in particular situations.

— *Tasks and scheduling models*. Used in project management.

— *Process and (multi)physics models*. Support engineering and simulation at plant process and plant system level. They ensure that the effects of the specified I&C functions are those expected at plant process or plant system level.

4.4.2. Model based systems engineering

INCOSE TP-2004-004-02 [51] states that:

"Model-based systems engineering (MBSE) is the formalized application of modelling to support system requirements, design, analysis, verification and validation activities beginning in the conceptual design phase and continuing throughout development and later life cycle phases."

MBSE is a branch of systems engineering that uses models as a means of information exchange between engineers, complementing the classical document based information exchange. One advantage of model based information exchange is improved precision and better comprehension, at least for those familiar with the modelling languages used (see Section 4.4.5).

However, for systems as large and complex as nuclear facilities, there is a pressing need to go beyond mere information exchange and to provide extensive tool support to assist the numerous and varied engineering activities during the system's life cycle. This includes:

— Functional validation of requirements, i.e. ensuring that the specified requirements are appropriate for the normal and abnormal situations the system might encounter.

— Step by step verification that contemplated solutions satisfy requirements in all situations.

— Failure analyses, such as failure modes, effects and criticality analysis and system theoretical process analysis, to ensure that there are appropriate defences against failures that could lead to unacceptable consequences.

— 'X in the loop' verification of implementations, with X being models, software, hardware or systems.

— Probabilistic safety or dependability analyses.

— Impact analysis of new engineering decisions.

— Exploration of models to help understand large and complex models.

— Optimization of design, construction, operation, outages, and decommissioning and deconstruction.

— Aids for operation during normal and abnormal situations and during maintenance or outages.

— Training, i.e. generation of training scenarios and verification of trainee actions.

More recently, MBSE has been used to address issues related to exploitation of models, using the following techniques:

— *Simulation*. The examination of the behaviour of dynamic phenomena models through simulation can be used to provide and verify evidence that a solution satisfies its requirements to better understand underlying mechanisms before developing appropriate solutions, or to support decision making, training and education. This is known as modelling and simulation based systems engineering (M&SBSE). Modelling and simulation allow digital experiments in well controlled and repeatable conditions, and in conditions that might be impractical or even unacceptable in physical experiments (e.g. experiments too long to allow the examination of many or even a single case, or severe accident conditions). Also, digital experiments are designed to be observable, as opposed to physical experiments where some phenomena might be too rapid or observation measurements deficient. M&S can be used for the prediction of behaviour and performance of systems, the evaluation of alternative solutions, the search for optimal solutions, the conduct of sensitivity analyses, or the support for HFE and ergonomics studies.

— *Formal analysis*. The application of mathematically rigorous techniques for the systematic verification of logical or quantitative properties. However, due to theoretical and practical limits, it is not always applicable.

— *Data validation and reconciliation*. Use of mathematical methods and models (preferably those developed during design) to correct measurement errors during operation (which could be due to inappropriate response times, lack of precision, miscalibration or failures), and to reduce margins of uncertainty.

— *Data assimilation*. A combination of models (also those developed during design) with observation data during operation to determine the most likely current state of the system, to interpolate limited observation data, to determine initial conditions for forecast ('What-if') models, to determine the causal factors that led to the current system state and to determine model parameters based on observed data.

These techniques require the use of formal modelling languages (i.e. languages with well defined syntax and semantics). MBSE also covers aspects related to the co-exploitation of models, where models of different types, from different sources and/or from different engineering disciplines are jointly used to ensure proper coordination between different project teams or with component suppliers. For example, co-simulation of a process physics model (e.g. a thermohydraulic model computing pressures, flows and temperatures in a plant system) and an I&C functional model may be used to verify that the specified I&C functions will contribute to the satisfaction of facility level requirements (e.g. that pressure and temperatures remain within acceptable limits).

The current understanding in modern systems engineering approaches is that although models and MBSE are the focal point, there is more to systems engineering than the models themselves. Models are represented by data and information, as are other systems engineering work products. In this context, models and other "work products are either projections of the same data and information or represented by data and information generated from other SE life cycle process activities" (see Ref. [52]). To manage increasingly complex systems of the future, there are benefits to managing this underlying data and information in such a way that it can be integrated and shared across the system development life cycle process activities, between the various systems engineering tools used to create and manage this data and information, and between organizations involved in the development and operation of the system of interest. This sharing will help ensure correctness, consistency and completeness of the data and information typical of the increasingly complex systems (see Ref. [53]).

As Ref. [54] states, the

"…information management process is a set of activities associated with the collection and management of information from one or more sources and the distribution of that information to one or more audiences. Information, in its most restricted technical sense, is an ordered sequence of symbols that record or transmit a message. The key idea is that information is a collection of facts that is organized in such a way that they have additional value beyond the value of the facts themselves. The systems engineer is both the generator and recipient of information products; thus, the systems engineer has a vital stake in the success of the development and use of the IM process and IM systems."

Each of the main processes of systems engineering (technical, technical management processes, agreement and organizational project enabling processes) have inputs, activities, controls, enablers and outputs. The inputs, controls and enablers for any given process are outputs of the activities of other processes, some internal to a project/organization and some external. The outputs or artefacts of any process are called work products with their underlying data and information. "These work products may be represented in a 'hard copy' printed form (documents, drawings, diagrams, etc.) or in an electronic form (documents, drawings, diagrams, databases, models, spreadsheets, etc.). In some cases, the electronic form may be a file without any underlying data or may be represented by underlying data and information stored in a database" [52].

Practising systems engineering from a datacentric perspective requires the electronic form of work products to be such that their underlying data and information are represented by a data set that can be shared and ideally integrated with other similarly formatted sets of data that adhere to industry interoperability standards. This allows the project to develop integrated, shareable sets of data from which the various work products across all life cycle process activities can be visualized.

Fundamental to the management of information is the management of data. Data management (DM) is a function that consists of the planning and execution of policies, practices and projects that acquire, control, protect, deliver and enhance the value of data and information assets (see Ref. [55]). The mission of the data management function is to meet and exceed the information needs of all stakeholders in the enterprise in terms of information availability, security and quality. To achieve this mission, the data management function has the following strategic goals. Data management experts need to understand the information requirements of the enterprise and all its stakeholders. As its main purpose, the data management process captures, stores, protects and ensures the integrity of the data assets. Additionally, it tries to continually improve the quality of data and information, including data accuracy, data integrity and data integration, the timeliness of data capture and presentation, the relevance and usefulness of data and the clarity and shared acceptance of data definitions.

4.4.3. Early application of modelling and simulation based systems engineering

Modelling and simulation (M&S) may be used during the system's life cycle, including in the very early stages. Examples of such application can be found in Ref. [56]. During preconceptual stages, M&S is useful at the level of the overall system. In the case of a nuclear facility, key stakeholders state their expectations and rationales regarding the contemplated facility and identify potential areas of risk. Modelling and simulation can support decision making, e.g. by providing information on the effects of different facility capability options on the grid (considering other likely changes such as massive introduction of renewables or widespread use of electric vehicles), or on the effects of different facility capability and construction schedule options on costs and revenues. At that stage, even if general principles are established regarding the I&C system, it is usually not sufficiently characterized to allow meaningful M&S.

During conceptual stages, M&S may be used to verify that the overall plant architecture supports the specified requirements, to support feasibility and sizing studies, or to prepare the safety justification of key innovative features.

4.4.4. Application of modelling and simulation based systems engineering for requirement engineering

Requirements specification is an essential activity in systems engineering. Requirements engineering is often defined as the process of defining, elaborating, documenting and maintaining requirements. Although these tasks are necessary, they are not sufficient: it needs to be ensured that the specified requirements and, in particular, functional and timing requirements, avoid different types of defects (see guidelines in Section 4.1.1). In particular, M&SBSE can be particularly useful to avoid the following:

— Inadequacy, which occurs when functional requirements are not appropriate for all situations the system may face. Situations can result from combinations of normal and abnormal states of the system, of the various elements constituting its environment, and from the operational goals assigned to it at any given time. As nuclear facilities rely increasingly on the flexibility and virtually unlimited functional capabilities of digital I&C systems, the number and ambition of functional requirements have soared. Experience shows that combined with the very large number of possible situations, this sometimes leads to functional requirements that fail to address certain situations or that are not fully adequate, even for safety systems. For example, the COMPSIS project report of the OECD Nuclear Energy Agency (OECD/NEA) [57] states:

"Weaknesses in requirements are one of the most significant contributors to systems and software failing to meet the intended goals. A better analysis is needed to understand the software's interfaces with the rest of the system and discrepancies between the documented requirements for a correct functioning system."

This is a serious issue, since inadequate functional requirements could defeat design diversity.

— Ambiguity, which occurs when requirements can be understood differently by different stakeholders (e.g. the specifier and the designer), or when requirements are expressed in such a way that there is no objective satisfaction criterion. As natural languages are inherently ambiguous, the solution is to use deterministic formal languages, such as functional block diagrams for the specification of functional I&C requirements. However, most such languages cannot everything that needs to be specified (e.g. response times, accuracy or limits to failure probability) and also can lead to over specification.
— Over specification, when requirements express elements belonging to the solution rather than to the problem to be solved. This often leads to requirements that are more complex than necessary (and thus more difficult to validate) and hinders the identification of more optimal solutions.
— Contradiction, when two or more requirements cannot be jointly satisfied. Though that will eventually be revealed during the design and verification process, it could cause serious cost overruns and delays for a project.

Requirements engineering is often considered a branch of systems engineering separate from MBSE. This is not ideal. When designing a solution for a system with given requirements, the practice is to identify subsystems, specify how they interact, and place requirements on each of them. When applying M&SBSE to verify the solution against system requirements, it is necessary to include, and thus to formally model, the system and the subsystems requirements.

Overall, M&SBSE can provide powerful approaches to help avoid the four defects mentioned:

— *Inadequacy*. Simulation places the system of interest within its operational context (which includes the environment of the system, the assumptions regarding the environment and the key requirements that need to be satisfied). It could be applied to cover the various normal and abnormal situations that the system may face to make sure that the technical, operational or design requirements do not conflict with the key requirements.
— *Ambiguity*. The use of modelling languages with well defined syntax and semantics significantly reduces the potential for ambiguity, and simulation may be used to 'animate' models and show their meaning.
— *Over specification*. As seen previously, deterministic formal languages, i.e. languages that, given initial and boundary conditions, determine a single, well defined behaviour, are not the best choices

for requirements modelling as they generally lead to over specification. Better choices are constraints based formal languages, i.e. languages that define envelopes (for timing and values) of acceptable or expected behaviours (see Fig. 6).

— *Contradiction.* M&SBSE applied to requirements engineering (especially using the formal language and practices presented in Section 4.3.1) can help reveal conflicting requirements. In some cases, formal models can be analysed using techniques such as model checking that can automatically identify contradictions. For models that are too complex to be subjected to such techniques, massive simulation (with the exploration of a large number of cases) may be conducted.

The requirement engineering process needs to be described in business procedures, taking into consideration standards, guidelines and possible defects. Insufficient or incorrect requirements can compromise the safety of a system.

Catastrophic accidents have been caused by requirements that were inappropriate in situations not foreseen or not taken into full consideration. The Cranbrook Manoeuvre is one such accident. It occurred in February 1978 at the Cranbrook International Airport, British Columbia, Canada, as follows:

— Incorrect deployment of aircraft thrust reversers during flight having caused several accidents, there was a requirement that reversers have to be disengaged when the wheels are not on the ground.
— Due to a very light traffic load, Cranbrook Airport's air traffic control (ATC) was conducted remotely from Calgary.
— Up to 1 m of snow had fallen in Cranbrook, and more was still falling.
— When the pilots announced their impending arrival to Calgary ATC, a snowplough was sent to clear the runway.
— Having taken a shorter route than expected, the aircraft initiated the landing procedure earlier than estimated by ATC. When the wheels touched the ground, the thrust reversers were deployed.
— A few seconds later, the pilots saw the snowplough on the runway: they had not seen it at first due to poor visibility conditions.
— They immediately ordered the stowing of the reversers, pushed the throttles to maximum power and took off.
— When the wheels left the ground, one reverser was fully stowed, but not the other, and because it was now disengaged, stowing could not be completed.
— Though the aircraft managed to clear the snowplough, aerodynamic pressure redeployed the thrust reverser completely. As the pilots did not have enough time to perform the necessary actions, the aircraft crashed, killing 42 of the 49 people on board.
— While the accident was caused by a combination of factors, one factor was that although the requirement regarding thrust reversers being disengaged during flight was appropriate in most situations, it was completely inadequate in this case, where the aircraft was configured for landing and the pilots abruptly changed the operational goal to take off.

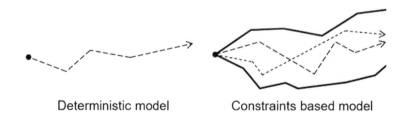

Deterministic model   Constraints based model

FIG. 6. Deterministic versus constraints models.

4.4.5. Application of modelling and simulation based systems engineering to nuclear facility I&C systems

Modelling for nuclear facility I&C systems can be carried out at the level of the overall I&C architecture or at the level of individual I&C systems and equipment. At both levels, it is possible to differentiate between requirements models and design, implementation and verification models.

Requirements models place the I&C architecture and the I&C systems within their overall context, and linked to other engineering disciplines that may place requirements on I&C, as listed in Section 2.5. They may cover a wide range of subjects: functions, interfaces, performance (e.g. response times or accuracy), reliability, safety, security and computer security, ambient conditions, power supplies, installation on-site, geometry, location in the facility, qualification, licensing and commissioning. Different requirements models may be used for different subjects, but as overall consistency needs to be ensured, not only for I&C but also for the rest of the facility, it is preferable to use a single requirements modelling framework for the entire facility. That framework could be tailored to the specific modelling needs of a given subject or group of related subjects using a specific metamodel agreed by the engineering teams involved and characterizing the elements and relationships used.

Metamodels facilitate mutual understanding for engineers. A metamodel describes the type of objects or classes and relationships used to build the models.

For example, an I&C architecture metamodel addressing safety and plant processes could have categories, such as initiating events, levels of defence in depth, I&C systems, I&C functions, data communication links, etc. I&C systems could be characterized by attributes such as their safety category or their technology. I&C functions could be characterized by attributes such as their safety category, their inputs and their outputs. Relationships between an event and some I&C functions could indicate which functions prevent the occurrence of the event or mitigate its effects, while others could indicate which I&C system implements individual functions.

Metamodels provide notation for design, implementation and verification models. As they constitute a concise and high level representation of the types of information in the various models, they facilitate the understanding of dependencies and complementarities between models.

Metamodels have to be described using formal notation. They can be described at a very fundamental level using, for example, meta object facility or can be extended from other languages like Unified Modelling Language (UML) or Systems Modelling Language (SysML).

Design, implementation and verification models for I&C systems are often (but not always) under the sole responsibility of I&C engineers, and thus could be specific to I&C engineering. However, as different subjects might need different types of models, one needs to ensure interoperability where necessary. Also, multiple organizations may contribute to I&C engineering (e.g. various I&C system or equipment suppliers, system integrators, safety or computer security assessors) and may have their own types of models. Here again, interoperability needs to be addressed.

Different models may be developed for I&C systems, serving different purposes as follows:

— Architectural and product breakdown models provide information about components (subsystems, cabinets, modules, software, hardware, documentation etc.) which may be used for manufacturing and configuration management, and for manufacturing and installation.
— Probabilistic or deterministic failure models may enrich architectural models to assess reliability and fault tolerance capability.
— Function block diagrams may be used to provide detailed, deterministic functional specifications in graphical form that could be simulated and verified against top level I&C requirements and then used for automatic or semiautomatic code generation.
— Interface models, and in particular data communication models and interoperability models, can be used to define and verify interactions between I&C systems and equipment. They may address issues such as communication protocols (in normal and failure conditions), mechanical and electrical or

optical interfaces, protection against failure propagation, communication bandwidth and loads in various conditions.

— Computer security models may be used to identify vulnerabilities and assess defensive measures.
— Allocation models may be used to determine an optimal number of controllers and to allocate I&C functions to individual controllers, based on function segregation rules (aiming at fault tolerance), on the resources necessary to each function (e.g. processing power, memory, input and output ports of various types, or communication bandwidth) and the resources each individual controller can provide.
— 3-D models may be used to place cabinets, cable paths and other I&C equipment in space, to verify that enough room is provided for installation, maintenance and replacement, and also to design control rooms that meet HFE requirements.
— HSI models may be used to verify that operator interfaces comply with HFE requirements.
— Thermal models may be used to determine the need for cooling and ventilation, so that I&C systems and equipment operate in the proper ambient conditions.

Sometimes these models can be combined into one, but the usual approach is to interconnect them.

4.4.6. Interlinkage with systems engineering processes

MBSE and models themselves are used in a range of applications and processes (see Sections 4.4.1 and 4.4.2):

— Stakeholder needs and requirements definition process or systems requirement definition process. The models are used to define, elaborate and trace requirements.
— Architecture and design definition processes. The models are used to describe overall I&C and individual I&C systems, their features and behaviours.
— Verification and validation processes. The models are used to verify I&C systems against applicable requirements.

The scope of the processes is not limited to those mentioned above. Models can be used in almost every systems engineering process and can be used to define processes themselves.

4.5. JUSTIFICATION FRAMEWORKS

Safety justification frameworks structure and organize the complex chains of reasoning justifying that a system (e.g. a nuclear facility, a plant system or an I&C system) complies with high level safety requirements, and also to help understand how that reasoning is supported by factual pieces of evidence (see Refs [58, 59]). Safety justification frameworks can support several systems engineering processes like verification and validation or information management.

Although their initial objective is to justify safety, they can also be used to justify compliance with any type of requirement, including non-safety requirements, and to show how combined compliance with component level requirements leads to compliance with system level requirements. They can also be used to justify the adequacy of requirements and the legitimacy of assumptions, and to explain the rationales of chosen solutions. They do so by integrating evidence from various sources, such as historical data, standards and regulations, design, operating procedures, methods and tools and thoroughness of verification. In the following, the term 'justification frameworks' is used to highlight this broadening of scope.

Justification frameworks are based on the three notions of claim, argument and evidence:

— A claim is an assertion that one seeks to justify. It is typically a statement about a property of a system, such as 'the system complies with requirement X', or 'the levels of defence in depth are adequately independent'. The claim can be treated as a requirement or an assumption which needs to be justified.
— Evidence is composed of the individual objective facts used in the justification of the claim. Sources of evidence include the design, development process, prior field experience, operation, testing and model or formal analysis.
— An argument provides explicit links between the claim and its pieces of evidence as illustrated in Fig. 7. The links are not necessarily tree like, and a subclaim (i.e. a claim supporting a part of the argument), or a piece of evidence may contribute to more than one claim.

There are different types of intermediate argument:

— Concretization is used when a claim, or some aspect of it, is given a more precise definition or interpretation. This may be the case for top level claims, which are sometimes expressed in abstract terms or in terms that cannot be achieved by real life systems (e.g. considering the inevitability of failures and finite response times and accuracy).
— Substitution is used when a claim about an object (or a property) is transformed into a claim about an equivalent object (or an equivalent property). For example, it is possible to substitute the claim that all components of a certain type have a certain property with a claim that a test specimen has that property. It can then be claimed that all production specimens also have this property, provided

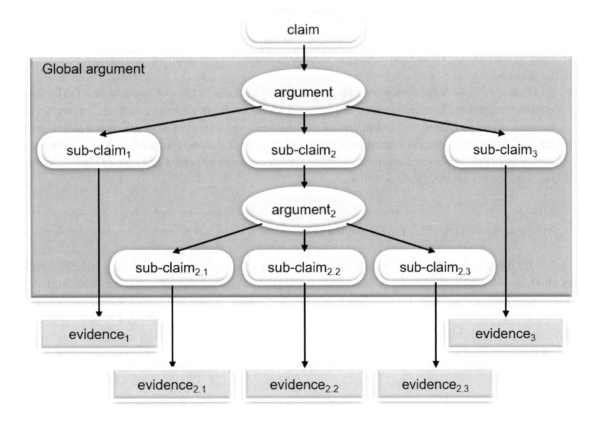

FIG. 7. Example of building a global argument.

that the claim can be justified that the specimens are equivalent in some clearly defined way to the test specimen.

— Decomposition is about partitioning some aspect of the claim (e.g. according to the functions or the architecture of the system, the properties being considered, or with respect to some sequence such as life cycle stages or modes of operation), provided that one can justify that together, the elements decomposition imply the initial claim.

— Calculation provides a quantitative argument when the value of one property of a system can be calculated from the values of other specific properties, possibly of other objects (e.g. subsystems).

In this way, a justification framework provides a structured link between a claim and the elements of its justification, which is more informative than simple traceability links. As it can integrate rigorous logic and formal modelling with human judgement, it also offers an insight into the reasoning that led to a chosen solution, which often helps identify possible weaknesses. This knowledge is also particularly useful to those who will have to operate, maintain and replace the system over multiple decades, when the original engineers are no longer available.

A justification framework may be used not only to link claims to existing evidence, but also be used very early in the system life cycle to outline the argument of top level claims before the availability of any actual piece of evidence. Explicit and structured argument outlines can help reasoning about possible solutions and about interactions and interfaces between solution elements. They may also be discussed among the teams, disciplines, stakeholders and organizations involved, and can thus facilitate communication and coordination.

The utility of justification frameworks is illustrated using the B737 MAX accident example introduced in Section 2.3:

— The classification of the consequences of an MCAS malfunction as 'major' was based on the assumption that pilots could reliably correct erroneous actions of the MCAS within 3 seconds. Such an assumption needs to be justified, and therefore is the object of a claim.

— A first argument step could be a decomposition listing the types of evidence that could collectively support the claim, such as tests in a flight simulator, human factors analyses and adequate provisions for training.

— Evidence for these subclaims would have to be provided later in the engineering process by other engineering teams. The decisions to not perform tests in a flight simulator and not informing and training pilots precluded the completion of the justification and would have raised warning signals. Tests in flight simulator performed after the second accident showed that with an MCAS maximum authority of 2.5 degrees and multiple spurious actuations, even informed and trained pilots could not reliably correct MCAS spurious actions.

4.6. MODEL BASED SAFETY ASSESSMENT

Model based safety assessment (MBSA) is a branch of systems engineering and MBSE. MBSA uses models for probabilistic safety assessments. However, safety assessments are not limited to probabilistic aspects and can benefit from M&SBSE. Indeed, there are cases where probabilistic parameters (e.g. failure probabilities) depend on deterministic operational conditions (e.g. temperature) and where pure probabilistic modelling needs to be closely associated with, or integrated into, other types of modelling.

4.7. MAINTENANCE OF ENGINEERING, SAFETY AND SECURITY KNOWLEDGE
DURING THE LIFETIME OF THE NUCLEAR FACILITY

As the lifetime of nuclear facilities spans several decades, sometimes 60 years or more, maintenance of engineering, safety and security knowledge is a necessity. It enables efficient, safe and secure plant operation. It also enables maintenance, retrofit, upgrade or replacement of plant components and systems.

In the case of I&C, and most particularly of digital I&C components and systems, long term maintenance of knowledge is a particularly critical issue. I&C has often been compared with a nervous system, as it contributes to, and controls the behaviour of, many — if not most — plant systems. Also, as the functional capability of digital technologies is virtually limitless, digital I&C systems play an increasing role in the efficiency and safety of modern nuclear facilities and their security, in particular their computer security, is an increasing concern.

MBSE, M&SBSE (including requirements engineering), MBSA and justification frameworks can be powerful tools for knowledge maintenance and for providing future generations with:

— Explicit and unambiguous statements of the operational context of each system (system boundaries, system environment, situations the system may face, interfaces with the environment, requirements and assumptions regarding the interfaces).
— Explicit and unambiguous specifications of requirements ('what' a system needs to achieve) and assumptions ('what' the system expects from its environment) in the different situations.
— Explicit and unambiguous descriptions of solutions ('how' a system satisfies its requirements) that can be simplified using simulation for better understanding.
— Explicit and unambiguous descriptions of system behaviour in an operational context.
— Explicit links between requirements (which need to be satisfied) and solution elements (how they contribute to the satisfaction of requirements).

Models can describe I&C systems using information items structured in hierarchy and interconnected, as described in Section 4.4.5. All changes in these models have to be implemented in a controllable manner and according to the configuration management process established in the project (see Section 3.3.2). Configuration baselines ensure consistency and integrity of the model throughout the life cycle.

MBSE helps to organize such information in a rigorous and formal way. The information is stored in the form of text, electronic tables or database on data media. However, both the data media and the format are prone to become obsolete and will need to be replaced. The information stored may be recovered in other forms if it is properly documented. In the case of MBSE, it is important to document the structure of the data model or metamodel. It helps to understand the full scope of objects that are stored in a database, including their attributes and the relation between them.

5. TOOLS FOR SYSTEMS ENGINEERING

Systems engineering processes cover different activities and deal with large amounts of information. Using an information management system (IMS) to manage these activities and data would improve the systems engineering process.

IMS supports the entire systems engineering life cycle for a given system. It represents an integrated suite of tools to cover all the activities in systems engineering. The following tools are typically part of the IMS supporting the systems engineering processes:

— Tools supporting organizational project enabling processes:
  - Graphical user interfaces, web clients;
  - Office automation, schedule, resources and financial management.
— Tools supporting technical processes:
  - Requirements management (including the design basis);
  - Piping and instrumentation diagrams;
  - I&C design (including but not limited to isometrics and layout);
  - Electrical system design;
  - Modelling (i.e. architecture, functionality);
  - Data capture, processing and validation.
— Tools supporting technical management processes:
  - Configuration and change management;
  - Document management;
  - Project execution control and outage planning;
  - Reporting;
  - Support for project reviews.
— Tools supporting regulatory/licensing processes.

For the selection of tools that make up the IMS, the following aspects may be considered:

— *Integration features with other tools.* As mentioned earlier, IMS is integrated with different tools that support specific activities. The tools need to have sufficient interfaces to ensure information exchange. In considering candidate tools, it is appropriate to check the availability of unified interfaces to support seamless integration.
— *Integration with other project stakeholders.* IMS receives different information from other participants, e.g. requirements, design bases. Moreover, it supports the transfer of information, e.g. the results of the design, configuration items, etc.
— *Access to data sources.* IMS ensures access to the same data sources for all stakeholders and disciplines in the project.
— *Scalability.* At the beginning of the project, it is very difficult to estimate how much information and how many activities are needed for systems engineering, and how many interfaces are to be managed. So, choosing an IMS that is scalable and flexible allows adaptability with any project.
— *Flexibility.* Many tools have inbuilt processes, e.g. configuration or change management. Often, a project has a specific feature of these processes, which requires adaptability by the IMS.
— *Tracking of the activities.* It is very important to track all the changes that have been implemented within the system.
— *Lifetime support.* The IMS is used during the entire life cycle, so it is important to choose reliable vendors and, updatable and upgradable technologies.
— *Computer security.* Interfaces within IMS and the role of the stakeholders involve different levels of information. The computer security aspects need to be considered in the IMS architecture and user rights.
— *HFE aspects.* The amount of data and their presentation require a user friendly concept to deliver the information requested.

With the support of the appropriate tools, transparency and efficiency can be increased, and information generated for one project can easily be transferred to other projects or project parts.

An IMS also facilitates the following life cycle processes in systems engineering:

— *Configuration baselines.* The storage and processing of configuration information, including systems, their parts with related documentation (e.g. requirements, specifications, design drawings,

V&V plans and reports, parts lists, test specifications, commissioning plans, maintenance manuals and operating handbooks).

— *The top–down approach for system function and structure.* The design bases, functional requirements, functional analysis, system architecture and equipment specifications and their interdependencies. The relationships need to be described to support V&V activities and data exchange.

— *System modelling.* Since I&C systems have many aspects (implementation of functions, HSI, real time characteristics, reliability, fault tolerance, etc.), modelling can decrease the risk of adverse influences on safety and of incurring costs later by defining additional requirements and by early detection of errors. This process is typically carried out only partially in the most basic design tools or using various specialized applications.

— *Procedures for life cycle activities.* Procedures typically cover interaction with other participants and define the responsibilities of each participant.

— *Information management and data exchange between stakeholders.* Many companies use different computerized technologies in their work. The common platform needs to be at least agreement of exchangeable configuration items attributes and communication protocols.

6. SUMMARY

This publication describes the philosophy and processes of systems engineering, based on ISO/IEC/IEEE 15288 [1]. As a well established consensus standard, systems engineering harmonizes hardware and software engineering into a system based methodology. This publication also provides insights for applying systems engineering methodologies for nuclear facilities and their I&C systems throughout the life cycle. It describes how systems engineering facilitates the implementation of digital technology into the nuclear facility with the potential to improve cost and shorten schedules for digital I&C projects. This publication introduces three major process areas for systems engineering:

— Organizational project enabling processes;
— Technical processes;
— Technical management process.

Each major systems engineering process area can be divided into a number of specific processes Detailed guidance is presented on applying each process to the nuclear facility and to I&C systems. The regulatory/licensing process is not identified in ISO/IEC/IEEE 15288 [1] and is included in this publication because of its impact on all aspects of the nuclear facility, especially I&C systems, and to support the integration of regulatory bodies as stakeholders within the systems engineering process.

A rigorous and well organized approach to developing new and modified digital I&C systems at nuclear facilities can avoid significant gaps in requirements and prevent unintended or undesirable behaviour that can be unsafe and/or extremely costly. This publication advocates a systems engineering approach to avoid such problems and manage technological complexity when designing the nuclear facility, the systems comprising the facility and specifically the facility's I&C systems.

Operational experience shows that challenges to effective design and life cycle management include coordinating inputs and products from numerous stakeholders as well as from teams in numerous engineering disciplines. Section 2.5 provides a list of some potential interfacing disciplines for consideration in applying systems engineering. In addition, Section 3.2.2 provides guidance on how to support effective stakeholder interactions. Section 4 discusses supporting methodologies, including the use of model based systems engineering as a means of information exchange between engineers to complement the classical document based information exchange. As the process of systems engineering matures, tools have been developed to assist in implementation. Section 5 identifies the use of information management systems as a tool to support the systems engineering processes.

Systems engineering is not just an I&C discipline but rather a holistic approach to designing systems throughout the nuclear facility. Although this publication focuses on the systems engineering approach for I&C systems, the nuclear facility would be well served to utilize this process for all systems.

Appendix

EXAMPLE PROCESSES FOR NUCLEAR FACILITY I&C SYSTEM DEVELOPMENT

Technical processes can be described in more detail in relation to specific life cycle stages or activities. They are used to describe, for example, the evolution or definition of configuration items during design processes or transformation and verification of requirements groups. Iterative processes and activities allow the management of design maturity.

The following processes are considered in this Appendix:

— *Functional structure development.* The definition of functions as well as functional requirements is considered part of the system requirements definition process (described in Section 1).
— *Definition of the overall I&C architecture.* This is considered part of the architecture definition processes (described in Section 3.2.4).
— *Development of workstations and control rooms.* This activity is considered part of the design definition process within the scope of I&C systems, but may require the involvement of experts from other disciplines (described in Section 3.2.5).
— *HSI development.* This is considered part of the design definition process within the scope of I&C, but may require the involvement of experts from other disciplines (described in Section 3.2.5).

The processes are described using business process modelling notation. Figure 8 shows the legend for the interpretation of the processes.

A.1. FUNCTIONAL STRUCTURE DEVELOPMENT PROCESS

The 'functional structure development' process aims at identifying monitoring and control functions. These functions are defined as result of elaboration and decomposition of process functions, derived from the functional analysis, and the requirements for the interactions between human and machine. The process combines the functional objectives and process functions hierarchies. It identifies and assigns the monitoring and control functions and their hierarchy within the overall I&C architecture, as well as to the human or machine.

This process:

— Performs analysis of monitoring and control tasks, including the tasks of I&C systems (control algorithms).
— Defines the functional HSI requirements, based on the specific characteristics of HSI components (shape, size, colour, etc.) for the video frames as well as of the panels.
— Defines I&C interfaces for the process equipment and nuclear facility operating personnel.

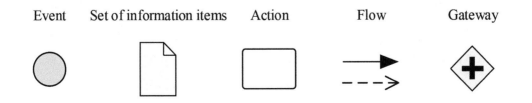

FIG. 8. Legend for the interpretation of processes.

— Determines the list of video frames, panels, functional specifications for control rooms and automated workstations based on monitoring and control tasks analysis.

Figure 9 presents the process workflow for producing those information items.

A.2. I&C ARCHITECTURE DEFINITION PROCESS

This process is focused on the development of the overall I&C architecture (Figs 10, 11). This architecture refers to a set of structural units (whose level of detail depends on the stage) and the interaction between them. As a result of these activities under this process, the I&C systems and their architecture are identified based on the input information (requirements, inputs of the 'functional structure development' process and others). The process of developing the overall I&C architecture is a step by step approach, and the initial version of the I&C architecture is defined based on the specific architecture requirements and requirements for defence in depth. This architecture is refined and updated in further steps according to the input data, including data from the I&C equipment.

The individual I&C system architecture is developed in a similar manner and follows a stage by stage process. It is based on the specific requirements of the I&C systems, the results of the overall I&C architecture design process and platform descriptions.

The requirements used in Figs 10 and 11 form part of the collection of all requirements allocated to I&C in terms of the overall I&C architecture and individual I&C systems.

A.3. DEVELOPMENT OF WORKSTATIONS AND CONTROL ROOM PROCESS

This process involves the development of the control rooms and arrangement of the workstations inside (Figs 12–14). Based on the input information (functional specifications, personnel, system descriptions, architecture of the building, ventilation design, communication facilities, etc.), the content and composition of the control room workstations are determined and solutions for functional zones, layout, illumination and room design are developed. The intermediate results of the process are proposals for adjusting the architectural plan, equipment layout or composition of workplaces, ceiling plan and other design solutions related to the control room as a complete system. The final result of the process is the design of the control room. The design is usually developed for the main control room, the emergency control room and the central control room. The specific nature of the control room is determined by design and depends on its importance for monitoring and control of the overall power unit.

A.4. HSI DEVELOPMENT PROCESS

The objective of this process is the development of the HSIs for the upper level system in control rooms, posts and local stations. Based on the input information (functional requirements for HSI, control room layout, operating experience, etc.), conceptual design solutions are developed. These solutions prevent human errors by providing the specific HSI structure, style guides and alarm solutions. Based on these solutions, the types of HSI, structure and layout principles of formats, panels and boards, and methods for the arrangement of secondary activities are developed. Once the suppliers of the HSI are identified, the video frames, panels and boards are designed and manufactured. The workflow for HSI development is presented in Figs 15–17.

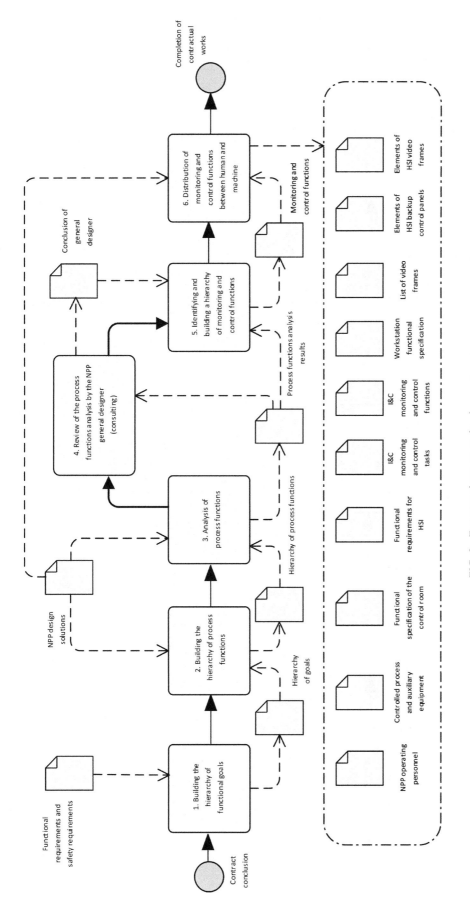

FIG. 9. Functional structure development process.

61

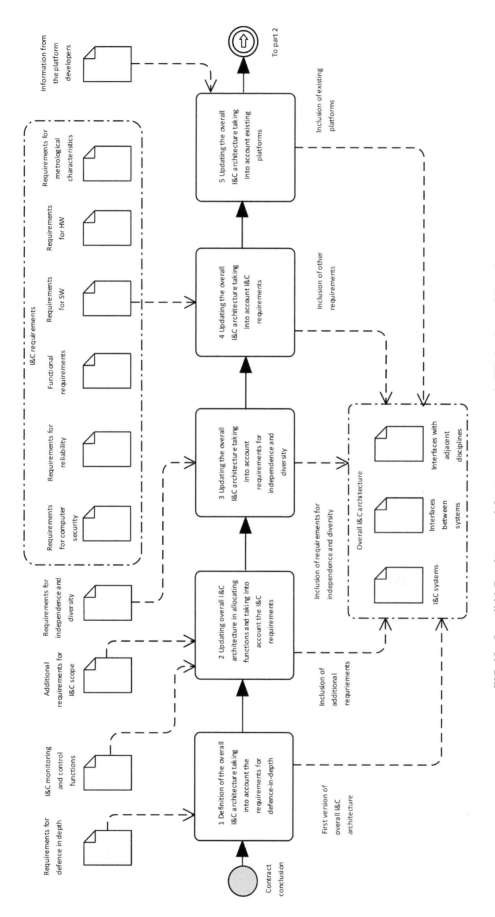

FIG. 10. Overall I&C architecture definition process — part 1 (HW: hardware; SW: software).

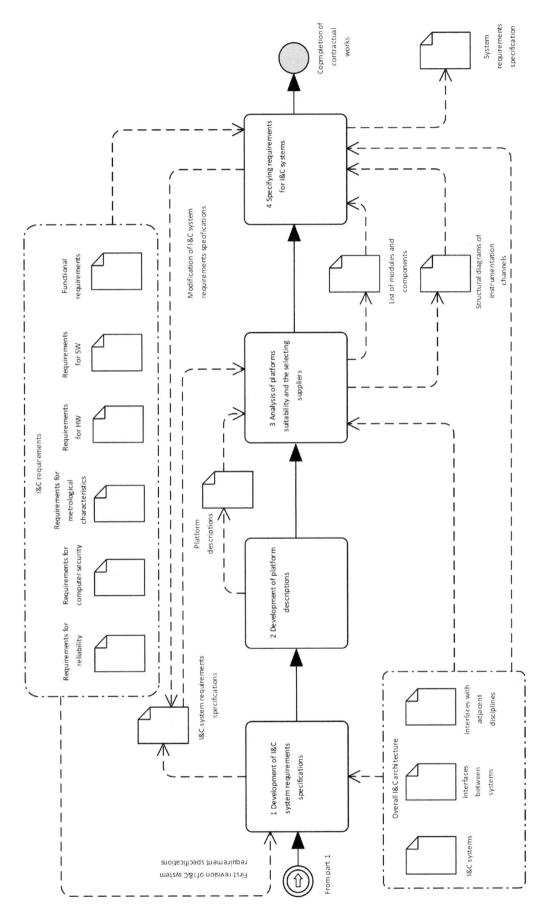

FIG. 11. Overall I&C architecture definition process — part 2.

63

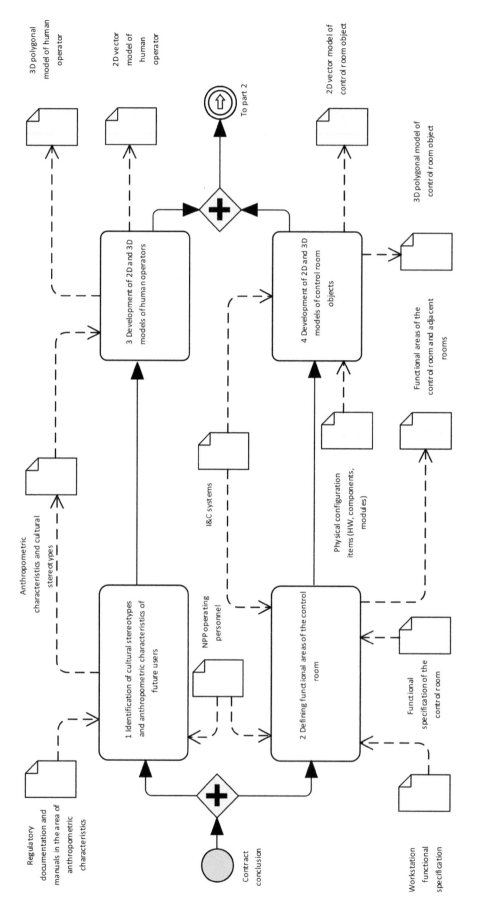

FIG. 12. *Development of workstations and control rooms process — part 1.*

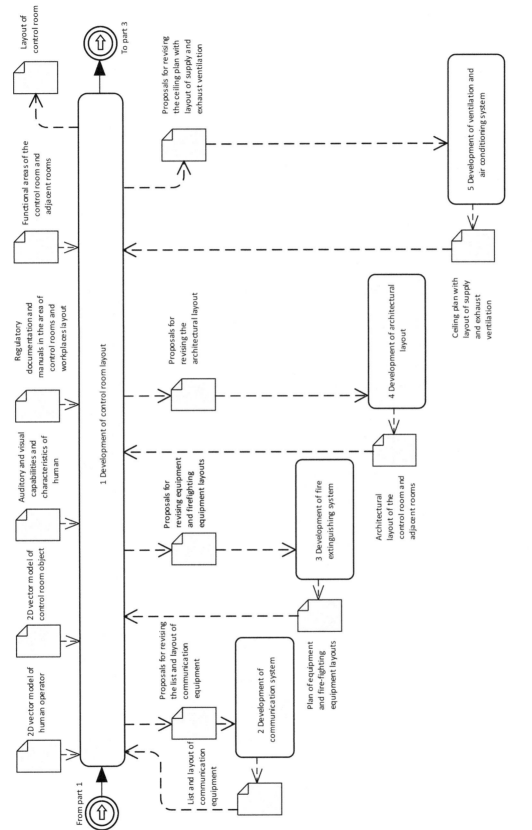

FIG. 13. *Development of workstations and control rooms process — part 2.*

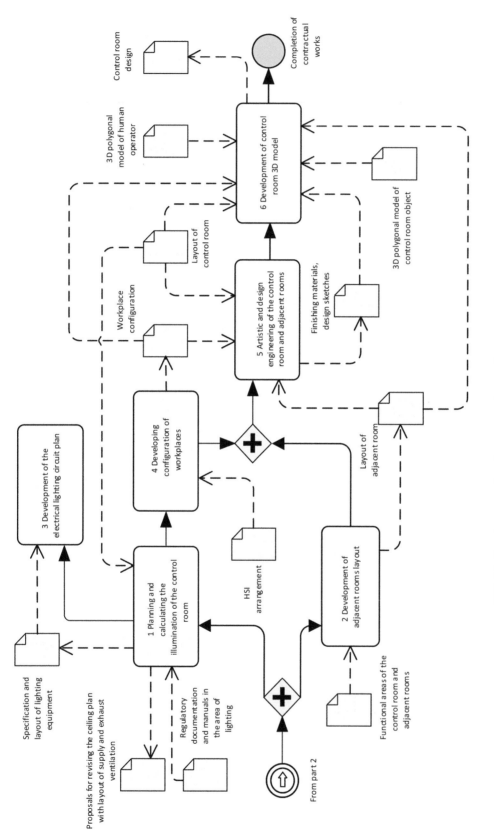

FIG. 14. Development of workstations and control rooms process — part 3.

66

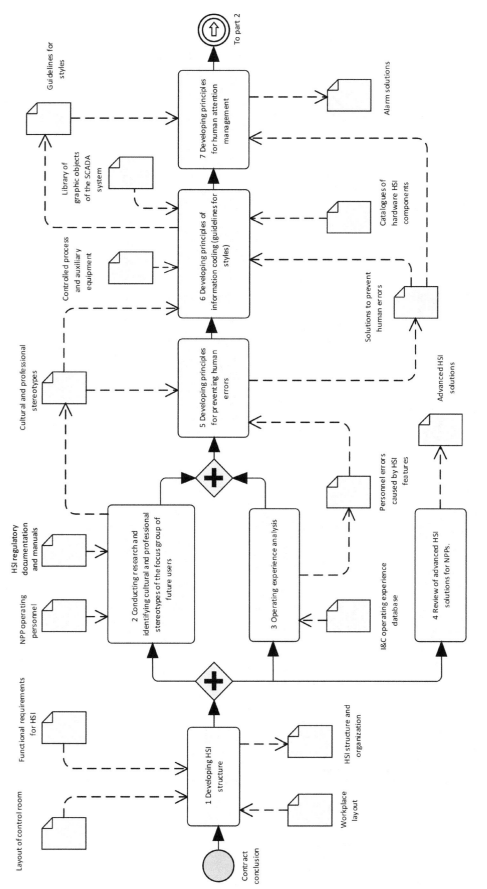

FIG. 15. HSI development process — part 1.

67

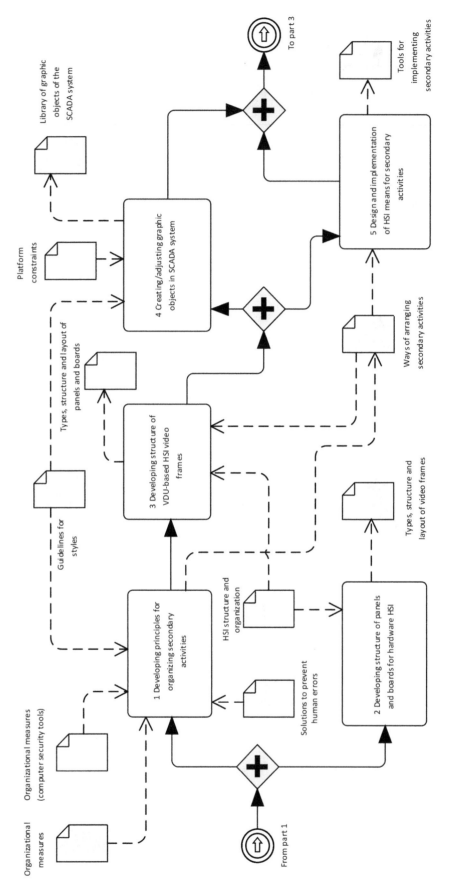

FIG. 16. HSI development process — part 2 (SCADA: supervisory control and data acquisition; VDU: visual display unit; HSI: human–system interface).

68

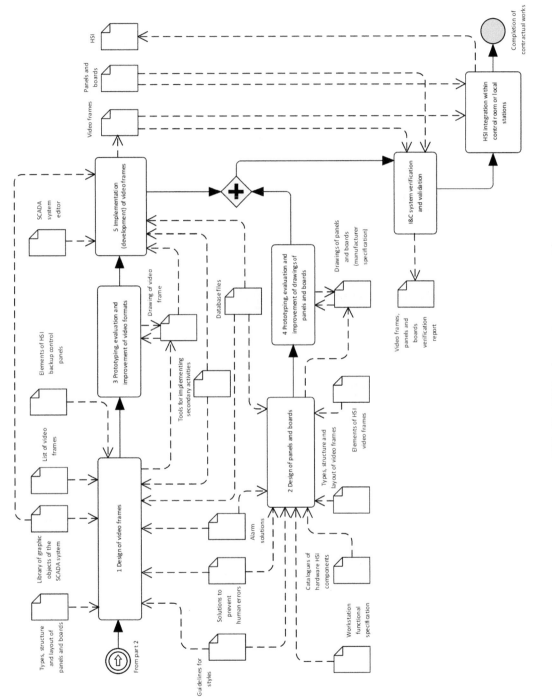

FIG. 17. HSI development process — part 3.

69

REFERENCES

[1] INTERNATIONAL ORGANIZATION FOR STANDARDIZATION, INTERNATIONAL ELECTROTECHNICAL COMMISSION, INSTITUTE OF ELECTRICAL AND ELECTRONICS ENGINEERS, Systems and Software Engineering — System Life Cycle Processes, ISO/IEC/IEEE 15288, ISO/IEC/IEEE, Geneva (2015).

[2] ELECTRIC POWER RESEARCH INSTITUTE, Digital Engineering Guide: Decision Making Using Systems Engineering, Rep. 3002011816, EPRI, Palo Alto, CA (2021).

[3] INDONESIAN TRANSPORTATION SAFETY BOARD, Aircraft Accident Investigation Report, Rep. KNKT.18.10.35.04, Komite Nasional Keselamatan Transportasi, Jakarta (2019).

[4] INTERNATIONAL ELECTROTECHNICAL COMMISSION, Nuclear Power Plants: Instrumentation and Control Systems Important to Safety — Classification of Instrumentation and Control Functions, IEC Standard 61226, 4th edn, IEC, Geneva (2020).

[5] INTERNATIONAL ATOMIC ENERGY AGENCY, Design of Instrumentation and Control Systems for Nuclear Power Plants, IAEA Safety Standards Series No. SSG-39, IAEA, Vienna (2016).

[6] INTERNATIONAL COUNCIL ON SYSTEMS ENGINEERING, Systems Engineering Handbook: A Guide for System Life Cycle Processes and Activities, 4th edn, INCOSE, New York (2015).

[7] INTERNATIONAL ELECTROTECHNICAL COMMISSION, Nuclear Power Plants: Instrumentation and Control Important to Safety — General Requirements for Systems, Rep. IEC 61513, IEC, Geneva (2011).

[8] INTERNATIONAL ATOMIC ENERGY AGENCY, Safety of Nuclear Power Plants: Design, IAEA Safety Standards Series No. SSR-2/1 (Rev. 1), IAEA, Vienna (2016).

[9] INTERNATIONAL ATOMIC ENERGY AGENCY, Computer Security Techniques for Nuclear Facilities, IAEA Nuclear Security Series No. 17-T (Rev. 1), IAEA, Vienna (2021).

[10] INTERNATIONAL ATOMIC ENERGY AGENCY, Computer Security of Instrumentation and Control Systems at Nuclear Facilities, IAEA Nuclear Security Series No. 33-T, IAEA, Vienna (2018).

[11] INTERNATIONAL ELECTROTECHNICAL COMMISSION, Nuclear Power Plants: Instrumentation and Control Systems Important to Safety: Software Aspects for Computer Based Systems Performing Category A Functions, IEC Standard 60880, IEC, Geneva (2006).

[12] INSTITUTE OF ELECTRICAL AND ELECTRONICS ENGINEERS, IEEE Standard Criteria for Programmable Digital Devices in Safety Systems of Nuclear Power Generating Stations, Rep. IEEE 7-4.3.2-2016, IEEE, New York (2016).

[13] INTERNATIONAL ELECTROTECHNICAL COMMISSION, Nuclear Power Plants: Instrumentation and Control Important to Safety — Hardware Design Requirements for Computer-Based Systems, Rep. IEC 60987:2007+AMD1:2013 CSV, IEC, Geneva (2013).

[14] INTERNATIONAL ELECTROTECHNICAL COMMISSION, Functional Safety of Electrical/ Electronic/Programmable Electronic Safety-Related Systems, IEC Standard 61508, IEC, Geneva (2010).

[15] INSTITUTE OF ELECTRICAL AND ELECTRONICS ENGINEERS, IEEE Standard Criteria for Safety Systems for Nuclear Power Generating Stations, IEEE Standard 603, IEEE, New York (2018).

[16] INTERNATIONAL ELECTROTECHNICAL COMMISSION, Nuclear Power Plants: Instrumentation and Control Systems Important to Safety — Software Aspects for Computer Based Systems Performing Category B or C Functions, IEC Standard 62138, IEC, Geneva (2018).

[17] INTERNATIONAL ELECTROTECHNICAL COMMISSION, Nuclear Power Plants: Instrumentation and Control Important to Safety — Development of HDL-Programmed Integrated Circuits for Systems Performing Category A Functions, IEC Standard 62566, IEC, Geneva (2012).

[18] INTERNATIONAL ELECTROTECHNICAL COMMISSION, Nuclear Power Plants: Instrumentation, Control and Electrical Power Systems — Cybersecurity Requirements, IEC Standard 62645, IEC, Geneva (2019).

[19] INTERNATIONAL ATOMIC ENERGY AGENCY, Core Knowledge on Instrumentation and Control Systems in Nuclear Power Plants, IAEA Nuclear Energy Series No. NP-T-3.12, IAEA, Vienna (2011).

[20] INTERNATIONAL ATOMIC ENERGY AGENCY, Stakeholder Involvement Throughout the Life Cycle of Nuclear Facilities, IAEA Nuclear Energy Series No. NG-T-1.4, IAEA, Vienna (2011).

[21] INTERNATIONAL ATOMIC ENERGY AGENCY, Human Factors Engineering in the Design of Nuclear Power Plants, IAEA Safety Standards Series No. SSG-51, IAEA, Vienna (2019).

[22] INTERNATIONAL ATOMIC ENERGY AGENCY, Human Factors Engineering Aspects of Instrumentation and Control System Design, IAEA Nuclear Energy Series No. NR-T-2.12, IAEA, Vienna (2021).

[23] INTERNATIONAL ELECTROTECHNICAL COMMISSION, Nuclear Power Plants: Control Rooms — Design, IEC Standard 60964, IEC, Geneva (2018).

[24] INTERNATIONAL ELECTROTECHNICAL COMMISSION, Nuclear Power Plants: Instrumentation and Control Systems — Requirements for Coordinating Safety and Cybersecurity, IEC Standard 62859, IEC, Geneva (2016).

[25] INTERNATIONAL ORGANIZATION FOR STANDARDIZATION, INTERNATIONAL ELECTROTECHNICAL COMMISSION, INSTITUTE OF ELECTRICAL AND ELECTRONICS ENGINEERS, Systems and Software Engineering — Life Cycle Processes — Requirements Engineering, ISO/IEC/IEEE 29148, ISO/IEC/IEEE, Geneva (2018).

[26] WESTERN EUROPEAN NUCLEAR REGULATORS' ASSOCIATION, Safety of New NPP Designs, Study by Reactor Harmonization Working Group RHWG, WENRA (2013).

[27] INTERNATIONAL ATOMIC ENERGY AGENCY, Approaches for Overall Instrumentation and Control Architectures of Nuclear Power Plants, IAEA Nuclear Energy Series No. NP-T-2.11, IAEA, Vienna (2018).

[28] INTERNATIONAL ATOMIC ENERGY AGENCY, Safety Classification of Structures, Systems and Components in Nuclear Power Plants, IAEA Safety Standards Series No. SSG-30, IAEA, Vienna (2014).

[29] INTERNATIONAL ELECTROTECHNICAL COMMISSION, Nuclear Power Plants: Instrumentation and Control Systems Important to Safety — Requirements for Coping with Common Cause Failure (CCF), IEC Standard 62340, IEC, Geneva (2007).

[30] INTERNATIONAL ELECTROTECHNICAL COMMISSION, Nuclear Power Plants: Instrumentation and Control Systems Important to Safety — Selection and Use of Industrial Digital Devices of Limited Functionality, IEC Standard 62671, IEC, Geneva (2013).

[31] NUCLEAR REGULATORY COMMISSION, Human Factors Engineering Program Review Model, Rep. NUREG-0711 (Rev. 3), Office of Nuclear Regulatory Research, NRC, Washington, DC (2012).

[32] INSTITUTE OF ELECTRICAL AND ELECTRONICS ENGINEERS, IEEE Recommended Practice for the Application of Human Factors Engineering to Systems, Equipment, and Facilities of Nuclear Power Generating Stations and Other Nuclear Facilities, IEEE Standard 1023-2004, IEEE, New York (2005).

[33] INSTITUTE OF ELECTRICAL AND ELECTRONICS ENGINEERS, IEEE Standard for System and Software Verification and Validation, IEEE Standard 1012, IEEE, New York (2016).

[34] INTERNATIONAL ATOMIC ENERGY AGENCY, Verification and Validation of Software Related to Nuclear Power Plant Instrumentation and Control, Technical Reports Series No. 384, IAEA, Vienna (1999).

[35] INTERNATIONAL ATOMIC ENERGY AGENCY, Application of Field Programmable Gate Arrays in Instrumentation and Control Systems of Nuclear Power Plants, IAEA Nuclear Energy Series No. NP-T-3.17, IAEA, Vienna (2016).

[36] INTERNATIONAL ATOMIC ENERGY AGENCY, Safety Culture in the Maintenance of Nuclear Power Plants, IAEA Safety Reports Series No. 42, IAEA, Vienna (2005).

[37] INTERNATIONAL ATOMIC ENERGY AGENCY, Maintenance, Surveillance and In-service Inspection in Nuclear Power Plants, IAEA Safety Standards Series No. NS-G-2.6, IAEA, Vienna (2002).

[38] INTERNATIONAL ATOMIC ENERGY AGENCY, Maintenance Optimization Programme for Nuclear Power Plants, IAEA Nuclear Energy Series No. NP-T-3.8, IAEA, Vienna (2018).

[39] INTERNATIONAL ATOMIC ENERGY AGENCY, Regulatory Surveillance of Safety Related Maintenance at Nuclear Power Plants, IAEA-TECDOC-960, IAEA, Vienna (1997).

[40] INTERNATIONAL ATOMIC ENERGY AGENCY, Advances in Safety Related Maintenance, IAEA-TECDOC-1138, IAEA, Vienna (2000).

[41] INTERNATIONAL ATOMIC ENERGY AGENCY, Configuration Management in Nuclear Power Plants, IAEA-TECDOC-1335, IAEA, Vienna (2003).

[42] INTERNATIONAL ATOMIC ENERGY AGENCY, Guidance for Optimizing Nuclear Power Plant Maintenance Programmes, IAEA-TECDOC-1383, IAEA, Vienna (2003).

[43] INTERNATIONAL ATOMIC ENERGY AGENCY, Operation and Maintenance of Spent Fuel Storage and Transportation Casks/Containers, IAEA-TECDOC-1532, IAEA, Vienna (2007).

[44] INTERNATIONAL ATOMIC ENERGY AGENCY, Application of Reliability Centred Maintenance to Optimize Operation and Maintenance in Nuclear Power Plants, IAEA-TECDOC-1590, IAEA, Vienna (2008).

[45] INTERNATIONAL ELECTROTECHNICAL COMMISSION, Nuclear Power Plants: Instrumentation and Control Important to Safety — Guidance for the Decision on Modernization, Rep. IEC TR 62096, IEC, Geneva (2009).

[46] INTERNATIONAL ATOMIC ENERGY AGENCY, Management of Life Cycle and Ageing at Nuclear Power Plants: Improved I&C Maintenance, IAEA-TECDOC-1402, IAEA, Vienna (2004).

[47] INTERNATIONAL ATOMIC ENERGY AGENCY, Application of Configuration Management in Nuclear Power Plants, Safety Reports Series No. 65, IAEA, Vienna (2010).

[48] INTERNATIONAL ATOMIC ENERGY AGENCY, Information Technology for Nuclear Power Plant Configuration Management, IAEA-TECDOC-1651, IAEA, Vienna (2010).

[49] INSTITUTE OF ELECTRICAL AND ELECTRONICS ENGINEERS, IEEE Standard for Software Configuration Management Plans, IEEE Standard 828, IEEE, New York (2012).

[50] HINCHEY, M.G., JONATHAN, J.P., Applications of Formal Methods, Prentice Hall International Series in Computer Science, Hoboken, NJ (1995).

[51] INTERNATIONAL COUNCIL ON SYSTEMS ENGINEERING, Systems Engineering Vision 2020, Rep. INCOSE-TP-2004-004-02, Rev. 2.03, INCOSE, San Diego, CA (2007).

[52] INTERNATIONAL COUNCIL ON SYSTEMS ENGINEERING, Integrated Data as a Foundation of Systems Engineering, Whitepaper by the Requirements Working Group, INCOSE, San Diego, CA (2018).

[53] WHEATCRAFT, L.S., RYAN, M.J., SVENSSON, C., Integrated Data as the Foundation of Systems Engineering. INCOSE International Symposium, Vol. 27 (2017) 1423–1437.

[54] INTERNATIONAL COUNCIL ON SYSTEMS ENGINEERING, SEBoK Editorial Board, Guide to the Systems Engineering Body of Knowledge (SEBoK), Wiki v. 2.3(2020),
https://www.sebokwiki.org/wiki/Guide_to_the_Systems_Engineering_Body_of_Knowledge_(SEBoK)

[55] DAMA INTERNATIONAL, DAMA-DMBOK: Data Management Body of Knowledge, 2nd Edition, Technics Publications, Sedona, AZ (2017).

[56] HAVEMAN, S.P., BONNEMA, G.M., Communication of Simulation and Modelling Activities in Early Systems Engineering, 2015 Conference on Systems Engineering Research, Procedia Computer Science, Vol. **44** (2015) 305–314.

[57] OECD NUCLEAR ENERGY AGENCY, Computer-Based Systems Important to Safety (COMPSIS) Project: Second Period Operation (2008–2011), Final Rep. NEA/CSNI/R(2012)12, OECD, Paris (2012).

[58] INTERNATIONAL ORGANIZATION FOR STANDARDIZATION, INTERNATIONAL ELECTROTECHNICAL COMMISSION, Systems and Software Engineering: Systems and Software Assurance — Part 2: Assurance Case, Standard 15026-2:2011, ISO/IEC, Geneva (2011).

[59] INTERNATIONAL ATOMIC ENERGY AGENCY, Dependability Assessment of Software for Safety Instrumentation and Control Systems at Nuclear Power Plants, IAEA Nuclear Energy Series No. NP-T-3.27, IAEA, Vienna (2018).

GLOSSARY

activity. Set of cohesive tasks of a process (ISO/IEC/IEEE 15288 [1]).

configuration management. The process of identifying and documenting the characteristics of a facility's structures, systems and components (including computer systems and software), and of ensuring that changes to these characteristics are properly developed, assessed, approved, issued, implemented, verified, recorded and incorporated into the facility documentation.

'Configuration' is used in the sense of the physical, functional and operational characteristics of the structures, systems and components and parts of a facility (IAEA Safety Glossary[1]).

enabling system. A system that supports a system of interest during its life cycle stages but does not necessarily contribute directly to its function during its operation (ISO/IEC/IEEE 15288 [1]).

functional requirements. Requirements that specify the required functions or behaviours of an item (IAEA Safety Standards Series No. SSG-39 [5]).

hazard. A source of potential harm or a situation with a potential for harm in terms of human injury, damage to health, property, or the environment, or some combination of these. (IEEE Standard 1012 [33]).

I&C architecture. Organizational structure of the instrumentation and control systems of the plant that are important to safety (IAEA Safety Standards Series No. SSG-39 [5]).

life cycle. Evolution of a system, product, service, project or other human-made entity from conception through retirement (ISO/IEC/IEEE 15288 [1]).

life cycle management. Life management (or lifetime management) in which due recognition is given to the fact that at all stages in the *lifetime* there may be effects that need to be taken into consideration (IAEA Safety Glossary[1]).

(1) An example is the approach to products, processes and services in which it is recognized that at all stages in the lifetime of a product (extraction and processing of raw materials, manufacturing, transport and distribution, use and reuse, and recycling and waste management) there are environmental impacts and economic consequences.
(2) The term 'life cycle' (as opposed to lifetime) implies that the life is genuinely cyclical (as in the case of recycling or reprocessing).

life cycle model. Framework of processes and activities concerned with the life cycle that may be organized into stages, which also acts as a common reference for communication and understanding (ISO/IEC/IEEE 15288 [1]).

process.
(1) A course of action or proceeding, especially a series of progressive stages in the manufacture of a product or some other operation.
(2) A set of interrelated or interacting activities that transforms inputs into outputs (ISO/IEC/IEEE 15288 [1]).

---

[1] INTERNATIONAL ATOMIC ENERGY AGENCY, IAEA Safety Glossary: Terminology Used in Nuclear Safety and Radiation Protection, 2018 Edition, IAEA, Vienna (2019).

A product is the result or output of a process (IAEA Safety Glossary[1]).

requirement. A statement which translates or expresses a need and its associated constraints and conditions (ISO/IEC/IEEE 15288 [1]).

safety life cycle. Necessary activities involved in the implementation of safety related systems occurring during a period of time that starts at the concept phase of a project and finishes when all safety related systems are no longer available for use (IEC 61513 [7] and IEC 61508 [14]).

— Note 1: The overall safety life cycle of the I&C causes the development of requirements for individual system safety life cycles.
— Note 2: The system safety life cycle refers to the activities of the overall I&C safety life cycle.

stage. Period within the life cycle of an entity that relates to the state of its description or realization (ISO/IEC/IEEE 15288 [1]).

stakeholder. A person or company with a concern or interest in the activities and performance of an organization, business, or system. The term 'stakeholder' is used in the same broad sense as interested party and the same provisos are necessary (as "interested party" in the IAEA Safety Glossary[1]).

system. A set of components which interact according to a design to perform a specific (active) function, in which an element of the system can be another system, called a subsystem. Examples are mechanical systems, electrical systems and instrumentation and control systems (IAEA Safety Glossary[1]).

system of interest. A system whose life cycle is under consideration in the context of applying systems engineering principles outlined in this publication (ISO/IEC/IEEE 15288 [1]).

systems engineering.
(1) Interdisciplinary approach governing the total technical and managerial effort required to transform a set of stakeholder needs, expectations and constraints into a solution and to support that solution throughout its life (ISO/IEC/IEEE 15288 [1]).
(2) Transdisciplinary and integrative approach to enable the successful realization, use and retirement of engineered systems using systems principles and concepts, and scientific, technological, and management method (INCOSE Systems Engineering Handbook [6]).

validation.
(1) The process of determining whether a product or service is adequate to perform its intended function satisfactorily. Validation (typically of a system) concerns checking against the specification of requirements, whereas verification (typically of a design specification, a test specification or a test report) relates to the outcome of a process.
(2) Confirmation by examination and by objective evidence that specified objectives have been met and specified requirements for a specific purpose and use or application have been fulfilled (IAEA Safety Glossary[1]).

verification.
(1) The process of determining whether the quality or performance of a product or service is as stated, as intended, or as required.
(2) Confirmation by examination and by objective evidence that specified objectives have been met and specified requirements for specific results have been fulfilled (IAEA Safety Glossary[1]).

ABBREVIATIONS

CM	configuration management
DM	data management
FAT	factory acceptance test
HDL	hardware description language
HFE	human factors engineering
HSI	human–system interface
HVAC	heating, ventilation, air-conditioning
I&C	instrumentation and control
IEC	International Electrotechnical Commission
IEEE	Institute of Electrical and Electronics Engineers
IM	information management
IMS	information management system
INCOSE	International Council on Systems Engineering
ISO	International Organization for Standardization
M&S	modelling and simulation
M&SBSE	modelling and simulation based systems engineering
MBSA	model based safety assessment
MBSE	model based systems engineering
MCAS	manoeuvring characteristics augmentation system
NF	nuclear facility
NPP	nuclear power plant
OECD	Organisation for Economic Co-operation and Development
RPS	reactor protection system
SE	systems engineering
SSC	structures, systems and components
SysML	Systems Modelling Language
TWG-NPPIC	Technical Working Group on Nuclear Power Plant Instrumentation and Control (IAEA)
UML	Unified Modelling Language
V&V	verification and validation
WENRA	Western European Nuclear Regulators' Association

CONTRIBUTORS TO DRAFTING AND REVIEW

Baek, S.M.	KEPCO E&C, Republic of Korea
Bartha, T.	Institute for Computer Science and Control, Hungary
Benchabane, J.L.	Framatome, Germany
Bi, Daowei	Shanghai Nuclear Engineering Research and Design Institute, China
Boring, R.L.	Idaho National Laboratory, United States of America
Chernyaev, A.	Rusatom Automated Control Systems, Russian Federation
Eiler, J.	International Atomic Energy Agency
Frost, S.	Office for Nuclear Regulation, United Kingdom
Gibson, M.	Electric Power Research Institute, United States of America
Kolchev, K.	Rusatom Automated Control Systems, Russian Federation
Miedl, H.	TÜV Rheinland Industrie Service GmbH, Germany
Nguyen, T.	Électricité de France, France
Odess-Gillett, W.	Westinghouse, United States of America
Pickelmann, J.	Framatome GmbH, Germany
Stattel, R.	Nuclear Regulatory Commission, United States of America
Turi, T.	Paks II Nuclear Power Plant, Hungary
Wendenkampf, E.	Framatome, Germany
Zorin, A.	Rusatom Automated Control Systems, Russian Federation

Technical Meeting

Vienna, Austria: 23–26 March 2021

Consultants Meetings

Vienna, Austria: 9–13 December 2019, 16–20 November 2020,
25–29 January 2021, 12–16 April 2021

Structure of the IAEA Nuclear Energy Series*

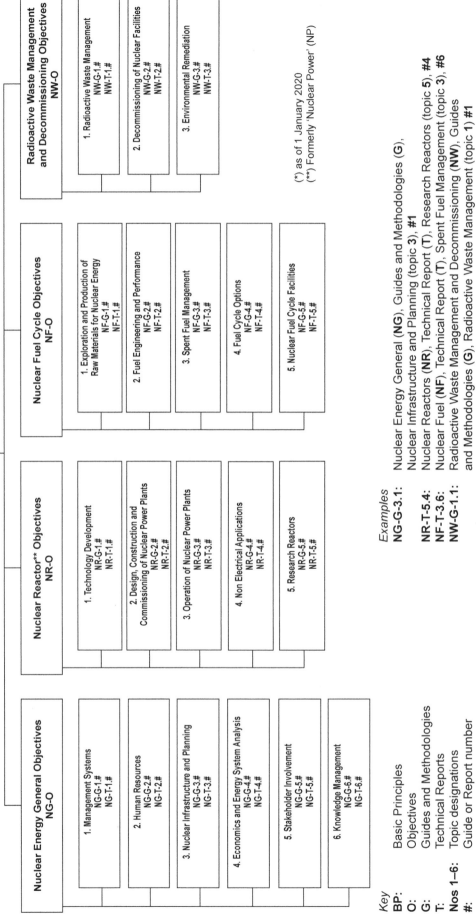

Nuclear Energy Basic Principles
NE-BP

Nuclear Energy General Objectives
NG-O

1. Management Systems
NG-G-1.#
NG-T-1.#

2. Human Resources
NG-G-2.#
NG-T-2.#

3. Nuclear Infrastructure and Planning
NG-G-3.#
NG-T-3.#

4. Economics and Energy System Analysis
NG-G-4.#
NG-T-4.#

5. Stakeholder Involvement
NG-G-5.#
NG-T-5.#

6. Knowledge Management
NG-G-6.#
NG-T-6.#

Nuclear Reactor Objectives**
NR-O

1. Technology Development
NR-G-1.#
NR-T-1.#

2. Design, Construction and Commissioning of Nuclear Power Plants
NR-G-2.#
NR-T-2.#

3. Operation of Nuclear Power Plants
NR-G-3.#
NR-T-3.#

4. Non Electrical Applications
NR-G-4.#
NR-T-4.#

5. Research Reactors
NR-G-5.#
NR-T-5.#

Nuclear Fuel Cycle Objectives
NF-O

1. Exploration and Production of Raw Materials for Nuclear Energy
NF-G-1.#
NF-T-1.#

2. Fuel Engineering and Performance
NF-G-2.#
NF-T-2.#

3. Spent Fuel Management
NF-G-3.#
NF-T-3.#

4. Fuel Cycle Options
NF-G-4.#
NF-T-4.#

5. Nuclear Fuel Cycle Facilities
NF-G-5.#
NF-T-5.#

Radioactive Waste Management and Decommissioning Objectives
NW-O

1. Radioactive Waste Management
NW-G-1.#
NW-T-1.#

2. Decommissioning of Nuclear Facilities
NW-G-2.#
NW-T-2.#

3. Environmental Remediation
NW-G-3.#
NW-T-3.#

(*) as of 1 January 2020
(**) Formerly 'Nuclear Power' (NP)

Key
BP: Basic Principles
O: Objectives
G: Guides and Methodologies
T: Technical Reports
Nos 1–6: Topic designations
#: Guide or Report number

Examples
NG-G-3.1: Nuclear Energy General (**NG**), Guides and Methodologies (**G**), Nuclear Infrastructure and Planning (topic **3**), **#1**
NR-T-5.4: Nuclear Reactors (**NR**), Technical Report (**T**), Research Reactors (topic **5**), **#4**
NF-T-3.6: Nuclear Fuel (**NF**), Technical Report (**T**), Spent Fuel Management (topic **3**), **#6**
NW-G-1.1: Radioactive Waste Management and Decommissioning (**NW**), Guides and Methodologies (**G**), Radioactive Waste Management (topic **1**) **#1**